Aylward and Findlay's

SI Chemical Data

7TH EDITION

BLACKMAN | GAHAN

WILEY

Seventh edition published 2014 by
John Wiley & Sons Australia, Ltd
42 McDougall Street, Milton, Qld 4064

First edition 1971
Second edition 1974
Third edition 1994
Fourth edition 1998
Fifth edition 2002
Sixth edition 2008

National Library of Australia
Cataloguing-in-Publication data

Author:	Blackman, Allan G., author.
Title:	Aylward and Findlay's SI chemical data/ Allan Blackman; Lawrence Gahan.
Edition:	7th edition.
ISBN:	9780730302469 (paperback)
Subjects:	Hazardous substances — Tables. Clinical chemistry — Tables. Chemical elements — Tables. Reference values (Medicine)
Other Authors/Contributors:	Gahan, Lawrence R., author. Aylward, G. H., author. SI chemical data 6th edition. Findlay, T. J. V. (Tristan John Victor), author. SI chemical data 6th edition.
Dewey Number:	540.212

Cover and internal design images: Michael Crawford

Printed in Singapore

M039886R15_290523

CONTENTS

ABOUT THE AUTHORS

Allan Blackman is an Associate Professor of Chemistry in the Department of Chemistry, University of Otago, Dunedin, New Zealand. He has a PhD in Physical Inorganic Chemistry, and his research interests include coordination chemistry and chemical kinetics. He has 22 years' experience in both teaching and research, is the co-author of a best-selling Australasian-focused first-year Chemistry textbook, and has won teaching awards at his home university. His monthly musings on things chemical can be found at http://neon.otago.ac.nz/chemistry/magazine/.

Lawrence R. Gahan is a Professor of Chemistry in the School of Chemistry and Molecular Biosciences, The University of Queensland, Australia. He has a PhD in Chemistry, has active research interests in bioinorganic chemistry, and has published extensively in international journals. Professor Gahan has 30 years' experience in university teaching with various teaching and learning grants, and local and national awards for his teaching. In addition, he has been awarded competitive grants to support his chemistry research.

PREFACE

The new edition of *SI Chemical Data* follows themes established by the previous authors, but incorporates extensive additions and changes. The adoption of the Globally Harmonised System of Classification and Labelling of Chemicals (GHS) has necessitated extensive revision of the hazard codes, which were elaborately detailed in earlier editions. Now, instead of individual hazard codes for each element or chemical, readers are advised to consult the many and detailed sources for the GHS and the abundance of material safety data sheets (MSDS) available from commercial and non-commercial sources. In addition to this major change, the order of the tables of data has been rearranged in an attempt to systematise the sequence of presentation. The values of the fundamental constants have been updated, data have been updated based on the most recent published compendia and, in some cases, data have been presented in different formats to those in previous editions. New tables have been included to provide data and information on:

- common radioisotopes
- the common amino acids
- the miscibility of common solvents
- ^{1}H chemical shifts of residual protons in deuterated NMR solvents
- ^{1}H and ^{13}C chemical shifts of common solvents
- common biological buffers
- the potentials and conversion factors for common reference electrodes.

An explanation of the GHS, with appropriate reference sources, has been included at the end of the book, along with information on the interpretation of an MSDS.

We gratefully acknowledge the extensive input of the many contributors to previous editions. In particular, we acknowledge that without the extraordinary labours of the original authors, Gordon Aylward and Tristan Findlay, in assembling and collating the data for the earlier editions of *SI Chemical Data*, this seventh edition would not have been possible.

A.G. Blackman
L.R. Gahan
July 2013

ACKNOWLEDGEMENTS

The authors and publisher would like to thank the following copyright holders, organisations and individuals for their permission to reproduce copyright material in *SI Chemical Data*, seventh edition.

Images

United Nations: **164,** UNECE GHS pictograms, © United Nations. Reproduced with permission. • Sigma-Aldrich Australia: **186**, Sigma-Aldrich Chemical Catalogue 2007–08, p. 2862 © Sigma-Aldrich Co. LLC. Used with permission.

Text

Armelle Vallat, Dr.: **123–24**, Adapted from 'Good Laboratory Practice for HPLC', by Armelle Vallat, May 2007, p. 5. Reproduced with permission from Armelle Vallat. • John Wiley & Sons, Inc.: **132**, Adapted from 'Spectrometric Identification of Organic Compounds 7th Edition' by Silverstein et al., 2005, Appendix G, p. 200; **137–38**, Adapted from 'Spectrometric Identification of Organic Compounds 7th Edition' by Silverstein et al., 2005, Appendix C, pp. 70–71. This material is reproduced with permission of John Wiley & Sons, Inc. • Copyright Clearance Center: **133–6**, Reprinted with permission from 'NMR Chemical Shifts of Common Laboratory Solvents as Trace Impurities' by Gottlieb, Kotlyar & Nudelman, *The Journal of Organic Chemistry*, Vol. 62, pp. 715–7515. Copyright 1997 American Chemical Society; **160**, Reprinted from *Inorganica Chimica Acta*, Vol. 298, Iss. 1, Pavlishchuk & Addison 'Conversion constants for redox potentials measured versus different reference electrodes in acetonitrile solutions at 25 °C', pp. 97–102 © 2000 with permission from Elsevier. • Sigma-Aldrich Australia: **163–65**, GHS Elements table and text from http://www.sigmaaldrich.com/safety-center/globally-harmonized.html © Sigma-Aldrich Co. LLC. Used with permission; **167–72**, Sigma-Aldrich Material Safety Data Sheet for Ethanol © Sigma-Aldrich Co. LLC. Used with permission.

Every effort has been made to trace the ownership of copyright material. Information that will enable the publisher to rectify any error or omission in subsequent editions will be welcome. In such cases, please contact the Permissions Section of John Wiley & Sons Australia, Ltd who will arrange for the payment of the usual fee.

1 THE INTERNATIONAL SYSTEM OF UNITS (SI)

SI base units

Physical quantity		Unit	
Name	Symbol	Name	Symbol
Length	l	metre	m
Mass	m	kilogram	kg
Time	t	second	s
Electric current	I	ampere	A
Thermodynamic temperature	T	kelvin	K
Luminous intensity	I_v	candela	cd
Amount of substance	n	mole	mol

Definitions of the SI base units (with date of most recent revision)

metre: the metre is the length of path travelled by light in vacuum during a time interval of 1/299 792 458 of a second. (1983)
kilogram: the kilogram is the unit of mass; it is equal to the mass of the international prototype of the kilogram. (1901)
second: the second is the duration of 9 192 631 770 periods of the radiation corresponding to the transition between the two hyperfine levels of the ground state of the ^{133}Cs atom. (1967)
ampere: the ampere is that constant current which, if maintained in two straight parallel conductors of infinite length, of negligible circular cross-section, and placed 1 metre apart in vacuum, would produce between these conductors a force equal to 2×10^{-7} newton per metre of length. (1948)
kelvin: the kelvin, unit of thermodynamic temperature, is the fraction 1/273.16 of the thermodynamic temperature of the triple point of water. (1967)
candela: the candela is the luminous intensity, in a given direction, of a source that emits monochromatic radiation of frequency 540×10^{12} hertz and that has a radiant intensity in that direction of 1/683 watt per steradian. (1979)
mole: the mole is the amount of substance of a system which contains as many elementary entities as there are carbon atoms in 0.012 kilogram of ^{12}C; its symbol is 'mol'. When the mole is used, the elementary entities must be specified and may be atoms, molecules, ions, electrons, other particles, or specified groups of such particles. (1971)

2 SI DERIVED UNITS COMMONLY USED IN CHEMISTRY

Measurement	Expression in terms of simpler quantities	Unit	Expression in terms of SI base units
area	length × width[a]	square metre	m^2
volume	length × width × height[a]	cubic metre	m^3
speed, velocity	distance[b]/time	metre per second	$m\,s^{-1}$
acceleration	velocity/time	metre per second squared	$m\,s^{-2}$
density	mass/volume	kilogram per cubic metre	$kg\,m^{-3}$
specific volume	volume/mass	cubic metre per kilogram	$m^3\,kg^{-1}$
force	mass × acceleration	newton, N	$1\,N = 1\,kg\,m\,s^{-2}$
pressure	force/area	pascal, Pa	$1\,Pa = 1\,N\,m^{-2} = 1\,kg\,m^{-1}\,s^{-2}$
energy	force × distance[b]	joule, J	$1\,J = 1\,N\,m = 1\,kg\,m^2\,s^{-2}$
power	energy/time	watt, W	$1\,W = 1\,J\,s^{-1} = 1\,kg\,m^2\,s^{-3}$
electric charge	electric current × time	coulomb, C	$1\,C = 1\,A\,s$
electric potential	energy/electric charge	volt, V	$1\,V = 1\,J\,C^{-1} = 1\,kg\,m^2\,s^{-3}\,A^{-1}$
electric resistance	electric potential/ electric current	ohm, Ω	$1\,\Omega = 1\,V/A = 1\,kg\,m^2\,s^{-3}\,A^{-2}$
electric conductance	electric current/ electric potential	siemens, S	$1\,S = 1\,A/V = 1\,kg^{-1}\,m^{-2}\,s^3\,A^2$
electric capacitance	electric charge/ electric potential	farad, F	$1\,F = 1\,C/V = 1\,kg^{-1}\,m^{-2}\,s^4\,A^2$

[a] Width and height are simply length in different directions.
[b] Distance is another name for length.

3 FUNDAMENTAL CONSTANTS (REVISED 2010)

The numbers in parentheses are the standard uncertainties in the last digits of the quoted value. Values that are given without associated uncertainties are exact, and therefore have no uncertainty.

Description	Symbol	Value
Avogadro constant	N_A	$6.022\ 141\ 29(27) \times 10^{23}$ mol^{-1}
Faraday constant	F	96 485.3365(21) C mol^{-1}
elementary charge	e	$1.602\ 176\ 565(35) \times 10^{-19}$ C
electron mass	m_e	$9.109\ 382\ 91(40) \times 10^{-31}$ kg
proton mass	m_p	$1.672\ 621\ 777(74) \times 10^{-27}$ kg
neutron mass	m_n	$1.674\ 927\ 351(74) \times 10^{-27}$ kg
atomic mass constant	m_u	$1.660\ 538\ 921(73) \times 10^{-27}$ kg
unified atomic mass unit	u	$1.660\ 538\ 921(73) \times 10^{-27}$ kg[a]
Planck constant	h	$6.626\ 069\ 57(29) \times 10^{-34}$ J s
Planck constant/2π	$\hbar$	$1.054\ 571\ 726(47) \times 10^{-34}$ J s
speed of light in vacuum	c_0	299 792 458 m s^{-1}
magnetic constant (permeability of a vacuum)	μ_0	$4\pi \times 10^{-7}$ N A^{-2}
electric constant (permittivity of a vacuum)	ε_0	$8.854\ 187\ 817 \times 10^{-12}$ F m^{-1}
Bohr radius	a_0	$5.291\ 772\ 109\ 2(17) \times 10^{-11}$ m
Hartree energy	E_h	$4.359\ 744\ 34(19) \times 10^{-18}$ J
Rydberg constant	R_∞	10 973 731.568 539(55) m^{-1}
Bohr magneton	μ_B	$9.274\ 009\ 68(20) \times 10^{-24}$ J T^{-1}
molar gas constant	R	8.314 462 1(75) J mol^{-1} K^{-1}
Boltzmann constant	k	$1.380\ 648\ 8(13) \times 10^{-23}$ J K^{-1}
molar volume of ideal gas (273.15 K, 100 kPa)[b]	V_m	$22.710\ 953(21) \times 10^{-3}$ m^3 mol^{-1}
molar volume of ideal gas (273.15 K, 101.325 kPa)[c]	V_m	$22.413\ 968(20) \times 10^{-3}$ m^3 mol^{-1}
standard acceleration of gravity	g_n	9.806 65 m s^{-2}

[a] m_u is equal to the unified atomic mass unit with symbol u; i.e. m_u = 1 u

[b] 100 kPa = 1 bar

[c] 101.325 kPa = 1 standard atmosphere

4 COMMON CONVERSION FACTORS

Energy

	joule	cal	erg	cm^3 atm	eV
1 joule	1	0.2390	10^7	9.869	6.242×10^{18}
1 calorie	4.184	1	4.184×10^7	41.29	2.612×10^{19}
1 erg	10^{-7}	2.390×10^{-8}	1	9.869×10^{-7}	6.242×10^{11}
1 cm^3 atm	0.1013	2.422×10^{-2}	1.013×10^6	1	6.325×10^{17}
1 eV	1.602×10^{-19}	3.829×10^{-20}	1.602×10^{-12}	1.581×10^{-18}	1

Equivalents of energy

	J mol^{-1}	cal mol^{-1}	erg molecule^{-1}
Wavenumber of 1 cm^{-1}	11.96	2.859	1.986×10^{-16}
Energy of 1 electronvolt (eV) per molecule	9.649×10^4	2.306×10^4	1.602×10^{-12}

Pressure

	pascal	atm	mmHg (Torr)	bar	dyne cm^{-2}	lbf in^{-2} (psi)
1 pascal	1	9.869×10^{-6}	7.501×10^{-3}	10^{-5}	10	1.450×10^{-4}
1 atm	1.013×10^5	1	760.0	1.013	1.013×10^6	14.70
1 mmHg (Torr)	133.3	1.316×10^{-3}	1	1.333×10^{-3}	1333	1.934×10^{-2}
1 bar	10^5	0.9869	750.1	1	10^6	14.50
1 dyne cm^{-2}	10^{-1}	9.869×10^{-7}	7.501×10^{-4}	10^{-6}	1	1.450×10^{-5}
1 lbf in^{-2} (psi)	6895	6.805×10^{-2}	51.71	6.895×10^{-2}	6.895×10^4	1

standard atmosphere (atm)	$1.013\ 25 \times 10^5$ Pa[a]
0 °C (ice point)	273.15 K[a]
litre (L)	1 dm^3 = 10^{-3} m^3
inch (in)	2.54×10^{-2} m[a]
pound (lb)	0.4536 kg

[a] These values are defined, and thus have zero uncertainty.

5 THE GREEK ALPHABET

Greek letter				Greek letter			
Upper-case	Lower-case	Greek name	English equivalent	Upper-case	Lower-case	Greek name	English equivalent
Α	α	alpha	a	Ν	ν	nu	n
Β	β	beta	b	Ξ	ξ	xi	x
Γ	γ	gamma	g	Ο	ο	omicron	o
Δ	δ	delta	d	Π	π	pi	p
Ε	ε	epsilon	e	Ρ	ρ	rho	r
Ζ	ζ	zeta	z	Σ	σ	sigma	s
Η	η	eta	e	Τ	τ	tau	t
Θ	θ	theta	th	Υ	υ	upsilon	u
Ι	ι	iota	i	Φ	ϕ	phi	ph
Κ	κ	kappa	k	Χ	χ	chi	ch
Λ	λ	lambda	l	Ψ	ψ	psi	ps
Μ	μ	mu	m	Ω	ω	omega	o

6 NUMERICAL PREFIXES

	Greek	Latin	IUPAC name of element[a]	IUPAC symbol of element[a]
0			nil	n
$\frac{1}{2}$	hemi	semi		
1	mono	uni	un	u
$1\frac{1}{2}$		sesqui		
2	di	bi	bi	b
3	tri	ter	tri	t
4	tetra	quadri	quad	q
5	penta	quinque	pent	p
6	hexa	sexi	hex	h
7	hepta	septi	sept	s
8	octa	octo	oct	o
9	ennea	nona	enn	e
10	deca	deci		
many	poly	multi		

[a] These are used to generate temporary names for newly discovered elements until an actual name is approved by IUPAC (the International Union of Pure and Applied Chemistry). For example, the temporary name of element 113 is ununtrium (Uut).

7 DECIMAL FRACTIONS AND MULTIPLES

Fraction	Prefix	Symbol	Multiple	Prefix	Symbol
10^{-1}	deci	d	10	deca	da
10^{-2}	centi	c	10^{2}	hecto	h
10^{-3}	milli	m	10^{3}	kilo	k
10^{-6}	micro	μ	10^{6}	mega	M
10^{-9}	nano	n	10^{9}	giga	G
10^{-12}	pico	p	10^{12}	tera	T
10^{-15}	femto	f	10^{15}	peta	P
10^{-18}	atto	a	10^{18}	exa	E
10^{-21}	zepto	z	10^{21}	zetta	Z
10^{-24}	yocto	y	10^{24}	yotta	Y

8 GROUND STATE ELECTRONIC CONFIGURATIONS OF THE ELEMENTS

n = principal quantum number
Z = atomic number
Note that the ground state electronic configurations of all elements following nobelium ($Z = 102$) are uncertain.

1–19

Shell n =		K 1	L 2		M 3			N 4				O 5				P 6				Q 7
Subshell		1*s*	2*s*	2*p*	3*s*	3*p*	3*d*	4*s*	4*p*	4*d*	4*f*	5*s*	5*p*	5*d*	5*f*	6*s*	6*p*	6*d*	6*f*	7*s*
Z	Element																			
1	H	1																		
2	He	2																		
3	Li	2	1																	
4	Be	2	2																	
5	B	2	2	1																
6	C	2	2	2																
7	N	2	2	3																
8	O	2	2	4																
9	F	2	2	5																
10	Ne	2	2	6																
11	Na	2	2	6	1															
12	Mg	2	2	6	2															
13	Al	2	2	6	2	1														
14	Si	2	2	6	2	2														
15	P	2	2	6	2	3														
16	S	2	2	6	2	4														
17	Cl	2	2	6	2	5														
18	Ar	2	2	6	2	6														
19	K	2	2	6	2	6		1												

8 GROUND STATE ELECTRONIC CONFIGURATIONS OF THE ELEMENTS *(continued)*

Shell		K	L		M			N				O				P				Q
$n =$		1	2		3			4				5				6				7
Subshell		1*s*	2*s*	2*p*	3*s*	3*p*	3*d*	4*s*	4*p*	4*d*	4*f*	5*s*	5*p*	5*d*	5*f*	6*s*	6*p*	6*d*	6*f*	7*s*
Z	Element																			
20	Ca	2	2	6	2	6		2												
21	Sc	2	2	6	2	6	1	2												
22	Ti	2	2	6	2	6	2	2												
23	V	2	2	6	2	6	3	2												
24	Cr	2	2	6	2	6	5	1												
25	Mn	2	2	6	2	6	5	2												
26	Fe	2	2	6	2	6	6	2												
27	Co	2	2	6	2	6	7	2												
28	Ni	2	2	6	2	6	8	2												
29	Cu	2	2	6	2	6	10	1												
30	Zn	2	2	6	2	6	10	2												
31	Ga	2	2	6	2	6	10	2	1											
32	Ge	2	2	6	2	6	10	2	2											
33	As	2	2	6	2	6	10	2	3											
34	Se	2	2	6	2	6	10	2	4											
35	Br	2	2	6	2	6	10	2	5											
36	Kr	2	2	6	2	6	10	2	6											
37	Rb	2	2	6	2	6	10	2	6			1								
38	Sr	2	2	6	2	6	10	2	6			2								
39	Y	2	2	6	2	6	10	2	6	1		2								
40	Zr	2	2	6	2	6	10	2	6	2		2								
41	Nb	2	2	6	2	6	10	2	6	4		1								
42	Mo	2	2	6	2	6	10	2	6	5		1								

Shell $n =$		K 1	L 2		M 3			N 4				O 5				P 6				Q 7
Subshell		1*s*	2*s*	2*p*	3*s*	3*p*	3*d*	4*s*	4*p*	4*d*	4*f*	5*s*	5*p*	5*d*	5*f*	6*s*	6*p*	6*d*	6*f*	7*s*
Z	Element																			
43	Tc	2	2	6	2	6	10	2	6	5		2								
44	Ru	2	2	6	2	6	10	2	6	7		1								
45	Rh	2	2	6	2	6	10	2	6	8		1								
46	Pd	2	2	6	2	6	10	2	6	10										
47	Ag	2	2	6	2	6	10	2	6	10		1								
48	Cd	2	2	6	2	6	10	2	6	10		2								
49	In	2	2	6	2	6	10	2	6	10		2	1							
50	Sn	2	2	6	2	6	10	2	6	10		2	2							
51	Sb	2	2	6	2	6	10	2	6	10		2	3							
52	Te	2	2	6	2	6	10	2	6	10		2	4							
53	I	2	2	6	2	6	10	2	6	10		2	5							
54	Xe	2	2	6	2	6	10	2	6	10		2	6							
55	Cs	2	2	6	2	6	10	2	6	10		2	6			1				
56	Ba	2	2	6	2	6	10	2	6	10		2	6			2				
57	La	2	2	6	2	6	10	2	6	10		2	6	1		2				
58	Ce	2	2	6	2	6	10	2	6	10	1	2	6	1		2				
59	Pr	2	2	6	2	6	10	2	6	10	3	2	6			2				
60	Nd	2	2	6	2	6	10	2	6	10	4	2	6			2				
61	Pm	2	2	6	2	6	10	2	6	10	5	2	6			2				
62	Sm	2	2	6	2	6	10	2	6	10	6	2	6			2				
63	Eu	2	2	6	2	6	10	2	6	10	7	2	6			2				
64	Gd	2	2	6	2	6	10	2	6	10	7	2	6	1		2				
65	Tb	2	2	6	2	6	10	2	6	10	9	2	6			2				

8 GROUND STATE ELECTRONIC CONFIGURATIONS OF THE ELEMENTS *(continued)* 66–88

Shell		K	L		M			N				O				P				Q
$n =$		1	2		3			4				5				6				7
Subshell		1*s*	2*s*	2*p*	3*s*	3*p*	3*d*	4*s*	4*p*	4*d*	4*f*	5*s*	5*p*	5*d*	5*f*	6*s*	6*p*	6*d*	6*f*	7*s*
Z	Element																			
66	Dy	2	2	6	2	6	10	2	6	10	10	2	6			2				
67	Ho	2	2	6	2	6	10	2	6	10	11	2	6			2				
68	Er	2	2	6	2	6	10	2	6	10	12	2	6			2				
69	Tm	2	2	6	2	6	10	2	6	10	13	2	6			2				
70	Yb	2	2	6	2	6	10	2	6	10	14	2	6			2				
71	Lu	2	2	6	2	6	10	2	6	10	14	2	6	1		2				
72	Hf	2	2	6	2	6	10	2	6	10	14	2	6	2		2				
73	Ta	2	2	6	2	6	10	2	6	10	14	2	6	3		2				
74	W	2	2	6	2	6	10	2	6	10	14	2	6	4		2				
75	Re	2	2	6	2	6	10	2	6	10	14	2	6	5		2				
76	Os	2	2	6	2	6	10	2	6	10	14	2	6	6		2				
77	Ir	2	2	6	2	6	10	2	6	10	14	2	6	7		2				
78	Pt	2	2	6	2	6	10	2	6	10	14	2	6	9		1				
79	Au	2	2	6	2	6	10	2	6	10	14	2	6	10		1				
80	Hg	2	2	6	2	6	10	2	6	10	14	2	6	10		2				
81	Tl	2	2	6	2	6	10	2	6	10	14	2	6	10		2	1			
82	Pb	2	2	6	2	6	10	2	6	10	14	2	6	10		2	2			
83	Bi	2	2	6	2	6	10	2	6	10	14	2	6	10		2	3			
84	Po	2	2	6	2	6	10	2	6	10	14	2	6	10		2	4			
85	At	2	2	6	2	6	10	2	6	10	14	2	6	10		2	5			
86	Rn	2	2	6	2	6	10	2	6	10	14	2	6	10		2	6			
87	Fr	2	2	6	2	6	10	2	6	10	14	2	6	10		2	6			1
88	Ra	2	2	6	2	6	10	2	6	10	14	2	6	10		2	6			2

Shell $n =$		K 1	L 2		M 3			N 4				O 5				P 6				Q 7
Subshell		1*s*	2*s*	2*p*	3*s*	3*p*	3*d*	4*s*	4*p*	4*d*	4*f*	5*s*	5*p*	5*d*	5*f*	6*s*	6*p*	6*d*	6*f*	7*s*
Z	Element																			
89	Ac	2	2	6	2	6	10	2	6	10	14	2	6	10		2	6	1		2
90	Th	2	2	6	2	6	10	2	6	10	14	2	6	10		2	6	2		2
91	Pa	2	2	6	2	6	10	2	6	10	14	2	6	10	2	2	6	1		2
92	U	2	2	6	2	6	10	2	6	10	14	2	6	10	3	2	6	1		2
93	Np	2	2	6	2	6	10	2	6	10	14	2	6	10	4	2	6	1		2
94	Pu	2	2	6	2	6	10	2	6	10	14	2	6	10	6	2	6			2
95	Am	2	2	6	2	6	10	2	6	10	14	2	6	10	7	2	6			2
96	Cm	2	2	6	2	6	10	2	6	10	14	2	6	10	7	2	6	1		2
97	Bk	2	2	6	2	6	10	2	6	10	14	2	6	10	8	2	6	1		2
98	Cf	2	2	6	2	6	10	2	6	10	14	2	6	10	10	2	6			2
99	Es	2	2	6	2	6	10	2	6	10	14	2	6	10	11	2	6			2
100	Fm	2	2	6	2	6	10	2	6	10	14	2	6	10	12	2	6			2
101	Md	2	2	6	2	6	10	2	6	10	14	2	6	10	13	2	6			2
102	No	2	2	6	2	6	10	2	6	10	14	2	6	10	14	2	6			2

9 PROPERTIES OF THE ELEMENTS

Table 9 contains names, symbols, physical properties, structures, metallic, covalent and ionic radii, and naturally occurring isotopes of the elements.

Z = atomic number.

M = molar mass (Wieser and Coplen 2010) based on $^{12}C = 12\ g\ mol^{-1}$ exactly. Molar mass values of all the elements with the exception of lithium (3) are known to at least 4 significant figures and are given to this precision in this table. For elements whose isotopes are all radioactive, the molar mass of the longest-lived radioactive isotope is shown in parentheses.

ρ (in $g\ cm^{-3}$) = density at 298 K. (l = liquid).

t_m (in °C) = melting temperature.

t_b (in °C) = boiling temperature (at 101.325 kPa).

d = decomposes; s = sublimes; * = melts under pressure; values in parentheses are estimated.

c_p (in $J\ K^{-1}\ g^{-1}$) = specific heat capacity at 298 K.

λ (in $J\ s^{-1}\ m^{-1}\ K^{-1}$) = thermal conductivity at 298 K. This is the energy transmitted through a unit cube per unit time when there is unit temperature difference between opposite sides.

κ (in $MS\ m^{-1}$) = electrical conductivity at 298 K. This is the electric current transmitted through a unit cube when there is unit potential difference between opposite sides.

Str. = structure of solid. Some of these structures are illustrated in table 17: 'Some crystal forms'.

- m = metal (coordination number in parentheses)
  - (12c) = cubic closest-packing (face-centred cubic)
  - (12h) = hexagonal closest-packing
  - (8) = body-centred cubic
  - (6c) = cubic
  - (6r) = rhombohedral
  - (6t) = tetragonal
- d = diamond
- 8 = layer
  - 8a = chain
- 9 = complicated
- – = structure unknown

Molecular structures are indicated by the appropriate formula.

r_{met} (in pm) = metallic radius (or atomic radius). This is one half of the shortest interatomic distance in the crystalline metal.

r_{cov} (in pm) = covalent radius. This is one half of the single bond interatomic distance in diatomic molecules of the elements, except where indicated.

r_{ion} (in pm) = ionic radius, for coordination number (CN) 6, except where otherwise indicated by a superscript, e.g. Cu 77(1+) is for Cu^+ with CN 6, but Cu $60(1+)^4$ is for Cu^+ with CN 4. Where not listed, values for CN other than 6 may be estimated from the following: ionic radii decrease by about 5% when the CN is decreased by 2, and increase by about 5% when the CN is increased by 2. Published values of ionic radii vary a great deal.

Only naturally occurring isotopes are listed, with per cent abundance in parentheses after the mass number. Radioactive isotopes are in italics.

Element	Symbol	Z	M / g mol^{-1}	ρ / g cm^{-3}	t_m / °C	t_b / °C	c_p / J K^{-1} g^{-1}	λ / J s^{-1} m^{-1} K^{-1}	κ / MS m^{-1}	*Str.*	r_{met} / pm	r_{cov} / pm	r_{ion} / pm	Isotopes Mass number (% abundance)
Actinium	Ac	89	(227.0)	10.1	1050	3200	0.12	12		m(12c)	188		112(3+)	
Aluminium	Al	13	26.98	2.70	660	2467	0.90	237	37	m(12c)	143		53(3+)	27(100)
Americium	Am	95	(243.1)	11.7	1176	(2067)	0.14	10	1.5	m(12h)	182		98(3+), 85(4+)	
Antimony	Sb	51	121.8	6.68	631	1635	0.21	24.3	2.3	m(6r)	145		76(3+), 60(5+)	121(57.2), 123(42.8)
Argon	Ar	18	39.95	1.40(l)	−189	−186	0.52	0.018		(12c)				36(0.3), 38(0.1), 40(99.6)
Arsenic														
(grey)	As	33	74.92	5.72	*817	s613	0.33	50	3.9	8	125		58(3+), 46(5+)	75(100)
(yellow)				2.03	d358					As_4		122(As_4)		
Astatine	At	85	(210.0)		302	(335)	0.14	1.7		–			62(7+)	
Barium	Ba	56	137.3	3.50	725	1640	0.20	18	2.9	m(8)	217		135(2+), 142(2+)^{8}	130(0.1), 132(0.1), 134(2.4), 135(6.6), 136(7.9), 137(11.2), 138(71.7)
Berkelium	Bk	97	(247.1)	15	986	(2623)				m			96(3+)	
Beryllium	Be	4	9.012	1.85	1278	2471	1.8	200	27	m(12h)	112		45(2+), 27(2+)^{4}	9(100)
Bismuth	Bi	83	209.0	9.8	271	1560	0.12	7.9	0.84	m(6r)	155		103(3+), 76(5+)	209(100)
Bohrium	Bh	107	[a]											
Boron	B	5	10.81[b]	2.34	2300	3660	1.03	27	10^{-10}	9		88(B_2Cl_4)	27(3+), 11(3+)^{4}	10(19.9), 11(80.1)
Bromine	Br	35	79.90	3.12	−7	59	0.47	0.12	10^{-16}	Br_2		114	196(1−)	79(50.7), 81(49.3)
Cadmium	Cd	48	112.4	8.65	321	767	0.23	97	15	m(12h)	149		95(2+), 110(2+)^{8}	106(1.3), 108(0.9), 110(12.5), 111(12.8), 112(24.1), *113*(12.2), 114(28.7), *116*(7.5)
Caesium	Cs	55	132.9	1.87	28	669	0.24	36	5.0	m(8)	265		167(1+), 174(1+)^{8}	133(100)
Calcium	Ca	20	40.08	1.55	842	1484	0.65	200	29	m(12c)	197		100(2+), 112(2+)^{8}	40(96.9), 42(0.6), 43(0.1), 44(2.1), 46(0.004), *48*(0.2)
Californium	Cf	98	(251.1)		900		0.10			m			95(3+)	

[a] The accurate molar mass has not yet been determined.
[b] See addendum at the end of this table.

Element	Symbol	Z	$\frac{M}{\text{g mol}^{-1}}$	$\frac{\rho}{\text{g cm}^{-3}}$	$\frac{t_m}{°C}$	$\frac{t_b}{°C}$	$\frac{c_p}{\text{J K}^{-1}\text{ g}^{-1}}$	$\frac{\lambda}{\text{J s}^{-1}\text{ m}^{-1}\text{ K}^{-1}}$	$\frac{\kappa}{\text{MS m}^{-1}}$	*Str.*	$\frac{r_{met}}{pm}$	$\frac{r_{cov}}{pm}$	$\frac{r_{ion}}{pm}$	Isotopes Mass number (% abundance)
Carbon (graphite)	C	6	12.01[b]	2.26	*3974	s3930	0.71	10[c] / 2000[d]	0.07	8		71(gr)	16(4+)	12(98.9), 13(1.1), *14*(trace)
(diamond)				3.51	>3550		0.51	1000	10^{-17}	d		77(d, C_2H_6)		
Cerium	Ce	58	140.1	6.78	795	3433	0.19	11	1.3	m(12c)	183		101(3+), 87(4+)	136(0.2), 138(0.2), 140(88.5), 142(11.1)
Chlorine	Cl	17	35.45[b]	1.56(l)	−101	−34	0.48	0.0089		Cl_2		99	181(1−)	35(75.8), 37(24.2)
Chromium	Cr	24	52.00	7.19	1857	2672	0.45	94	7.9	m(8)	125		80(2+), 62(3+), 44(6+)	50(4.3), 52(83.8), 53(9.5), 54(2.4)
Cobalt	Co	27	58.93	8.90	1495	2870	0.42	100	16	m(12h)	125		74(2+), 61(3+)	59(100)
Copernicium	Cn	112	(285.2)											
Copper	Cu	29	63.55	8.96	1085	2572	0.39	401	58.4	m(12c)	128		77(1+), 73(2+), 60(1+)[4]	63(69.2), 65(30.8)
Curium	Cm	96	(247.1)	13.5	1340	3110				m	174		97(3+)	
Darmstadtium	Ds	110	(281.2)											
Dubnium	Db	105	(268.1)											
Dysprosium	Dy	66	162.5	8.54	1410	2567	0.17	11	1.1	m(12h)	175		91(3+)	156(0.06), 158(0.1), 160(2.3), 161(18.9), 162(25.5), 163(24.9), 164(28.2)
Einsteinium	Es	99	(252.1)		(860)					–				
Erbium	Er	68	167.3	9.05	1529	2900	0.17	14	1.1	m(12h)	173		89(3+)	162(0.1), 164(1.6), 166(33.6), 167(23.0), 168(26.8), 170(14.9)
Europium	Eu	63	152.0	5.24	826	1596	0.18	14	1.1	m(8)	199		117(2+), 95(3+)	151(47.8), 153(52.2)
Fermium	Fm	100	(257.1)							–				
Flerovium	Fl	114	(289.2)											
Fluorine	F	9	19.00	1.51(l)	−220	−188	0.82	0.028		F_2		71	133(1−)	19(100)
Francium	Fr	87	(223.0)		(27)	(680)	0.37	15		m			180(1+)	
Gadolinium	Gd	64	157.3	7.90	1313	3273	0.24	10.6	0.75	m(12h)	179		94(3+)	152(0.2), 154(2.2), 155(14.8), 156(20.5), 157(15.6), 158(24.8), 160(21.9)

[b] See addendum at the end of this table.
[c] Conductivity in the direction perpendicular to the graphite sheet.
[d] Conductivity parallel to the graphite sheet.

Element	Symbol	Z	M / g mol^{-1}	ρ / g cm^{-3}	t_m / °C	t_b / °C	c_p / J K^{-1} g^{-1}	λ / J s^{-1} m^{-1} K^{-1}	κ / MS m^{-1}	*Str.*	r_{met} / pm	r_{cov} / pm	r_{ion} / pm	Isotopes Mass number (% abundance)
Gallium	Ga	31	69.72	5.91	30	2403	0.37	41	3.9	9	122		62(3+)	69(60.1), 71(39.9)
Germanium	Ge	32	72.63	5.32	937	2830	0.32	60	10^{-4}	d	123	121(Ge_2H_6)	73(2+), 53(4+)	70(20.8), 72(27.6), 73(7.7), 74(36.3), 76(7.6)
Gold	Au	79	197.0	19.3	1064	2856	0.13	317	44	m(12c)	144		137(1+), 85(3+)	197(100)
Hafnium	Hf	72	178.5	13.3	2233	4450	0.14	23	3.0	m(12h)	156		71(4+)	*174*(0.2), 176(5.2), 177(18.6), 178(27.3), 179(13.6), 180(35.1)
Hassium	Hs	108	(277.2)											
Helium	He	2	4.003	0.15(l)	*−272	−269	5.19	0.15		(12h)				3(10^{-4}), 4(100)
Holmium	Ho	67	164.9	8.80	1472	2694	0.16	16	1.2	m(12h)	174		90(3+)	165(100)
Hydrogen	H	1	1.008[b]	0.07(l)	−259	−253	14.3	0.18		H_2		37	208(1−)	1(99.99), 2(0.01)
Indium	In	49	114.8	7.31	157	2080	0.23	82	12	m(12c)	136		80(3+)	113(4.3), *115*(95.7)
Iodine	I	53	126.9	4.94	114	184	0.21	0.45	10^{-13}	I_2		133	220(1−)	127(100)
Iridium	Ir	77	192.2	22.5	2450	4500	0.13	147	19	m(12c)	136		68(4+)	191(37.3), 193(62.7)
Iron	Fe	26	55.85	7.86	1535	2750	0.45	80	10.3	m(8)	124		78(2+), 64(3+)	54(5.8), 56(91.8), 57(2.1), 58(0.3)
Krypton	Kr	36	83.80	2.16(l)	−157	−152	0.25	0.0095		(12c)		117(KrF_2)		78(0.3), 80(2.3), 82(11.6), 83(11.5), 84(57.0), 86(17.3)
Lanthanum	La	57	138.9	6.17	920	3470	0.19	14	1.8	m(12h)	187		103(3+)	*138*(0.1), 139(99.9)
Lawrencium	Lr	103	(262.1)							–				
Lead	Pb	82	207.2	11.3	327	1740	0.13	35	4.8	m(12c)	175		119(2+), 78(4+)	204(1.4), 206(24.1), 207(22.1), 208(52.4)
Lithium	Li	3	6.94[b]	0.53	180	1342	3.6	85	10.6	m(8)	152	134	76(1+), 59(1+)^{4}	6(7.5), 7(92.5)
Livermorium	Lv	116	[a]											
Lutetium	Lu	71	175.0	9.84	1663	3402	0.15	16	1.7	m(12h)	172		86(3+)	175(97.4), *176*(2.6)
Magnesium	Mg	12	24.31	1.74	650	1110	1.02	156	22	m(12h)	160		72(2+)	24(79.0), 25(10.0), 26(11.0)
Manganese	Mn	25	54.94	7.43	1244	2095	0.48	7.8	0.69	m(12c)	137		83(2+), 64(3+), 46(7+)	55(100)
Meitnerium	Mt	109	(276.2)											

[a] The accurate molar mass has not yet been determined.
[b] See addendum at the end of this table.

Element	Symbol	Z	M / g mol^{-1}	ρ / g cm^{-3}	t_m / °C	t_b / °C	c_p / J K^{-1} g^{-1}	λ / J s^{-1} m^{-1} K^{-1}	κ / MS m^{-1}	*Str.*	r_{met} / pm	r_{cov} / pm	r_{ion} / pm	Isotopes Mass number (% abundance)
Mendelevium	Md	101	(258.1)		827					–				
Mercury	Hg	80	200.6	13.53	−39	357	0.14	8.3	1.0	m(6r)	150	122(Hg_2F_2)	102(2+)	196(0.1), 198(10.0), 199(16.9), 200(23.1), 201(13.2), 202(29.8), 204(6.9)
Molybdenum	Mo	42	95.96	10.2	2610	5560	0.25	138	19	m(8)	136		69(3+), 65(4+), 59(6+)	92(14.8), 94(9.3), 95(15.9), 96(16.7), 97(9.6), 98(24.1), 100(9.6)
Neodymium	Nd	60	144.2	7.00	1024	3074	0.19	16	1.6	m(12h)	181		98(3+)	142(27.1), 143(12.2), *144*(23.8), 145(8.3), 146(17.2), 148(5.8), 150(5.6)
Neon	Ne	10	20.18	1.20(l)	−249	−246	1.03	0.049		(12c)				20(90.5), 21(0.3), 22(9.2)
Neptunium	Np	93	(237.0)	20.4	637	(3900)		6.3	0.82	9	130		101(3+), 87(4+)	
Nickel	Ni	28	58.69	8.90	1455	2730	0.44	91	14	m(12c)	125		69(2+), 60(3+)	58(68.1), 60(26.2), 61(1.2), 62(3.6), 64(0.9)
Niobium	Nb	41	92.91	8.57	2468	4758	0.27	54	6	m(8)	143		64(5+)	93(100)
Nitrogen	N	7	14.01[b]	0.81(l)	−210	−196	1.04	0.026		N_2		73(N_2H_4)	146(3−), 16(3+), 13(5+)	14(99.6), 15(0.4)
Nobelium	No	102	(259.1)							–			110(2+)	
Osmium	Os	76	190.2	22.6	3033	5012	0.13	88	11	m(12h)	134		63(4+)	184(0.02), *186*(1.6), 187(2.0), 188(13.2), 189(16.1), 190(26.3), 192(40.8)
Oxygen (ozone)	O	8	16.00[b]	1.15(l) 1.61(l)	−219 −193	−183 −111	0.92	0.026		O_2 O_3		74(H_2O_2)	140(2−)	16(99.8), 17(0.04), 18(0.2)
Palladium	Pd	46	106.4	12.0	1554	2970	0.24	72	9.3	m(12c)	138		86(2+)	102(1.0), 104(11.2), 105(22.3), 106(27.3), 108(26.5), 110(11.7)
Phosphorus (white) (red) (black)	 P	 15	 30.97	 1.82 2.35 2.7	 44 *590	 280 s417	 0.77 0.68 0.70	 0.24 12	 10^{-15}	 P_4 8a 8		 110(P_4)	 44(3+), 38(5+), 17(5+)[4]	 31(100)
Platinum	Pt	78	195.1	21.4	1772	3825	0.13	72	9.3	m(12c)	139		80(2+)	*190*(0.01), 192(0.8), 194(32.9), 195(33.8), 196(25.3), 198(7.2)
Plutonium	Pu	94	(244.1)	19.8	640	3230	0.13	6.7	0.7	9	151		100(3+), 86(4+)	

[b] See addendum at the end of this table.

Element	Symbol	Z	M / g mol^{-1}	ρ / g cm^{-3}	t_m / °C	t_b / °C	c_p / J K^{-1} g^{-1}	λ / J s^{-1} m^{-1} K^{-1}	κ / MS m^{-1}	*Str.*	r_{met} / pm	r_{cov} / pm	r_{ion} / pm	Isotopes Mass number (% abundance)
Polonium	Po	84	(209.0)	9.4	254	962	0.13	20	0.7	m(6c)	167		67(6+)	
Potassium	K	19	39.10	0.86	63	760	0.76	102	14	m(8)	227	196	138(1+)	39(93.3), *40*(0.01), 41(6.7)
Praseodymium	Pr	59	140.9	6.77	935	3510	0.19	12	1.5	m(12h)	132		99(3+), 85(4+)	141(100)
Promethium	Pm	61	(144.9)	7.22	1042	(2730)	0.19	18	1.6	m			97(3+)	
Protactinium	Pa	91	231.0	15.4	1572	(4000)	0.12	47	5.6	m	161		104(3+), 90(4+)	*231*(100)
Radium	Ra	88	(226.0)	5	700	1737	0.12	19	1.0	m			142(2+)[8]	*226* most abundant
Radon	Rn	86	(222.0)	4.4(l)	−71	−62	0.09	0.004		(12c)				
Roentgenium	Rg	111	(280.2)											
Rhenium	Re	75	186.2	21.0	3185	5596	0.14	48	5.2	m(12h)	137		63(4+), 53(7+)	185(37.4), *187*(62.6)
Rhodium	Rh	45	102.9	12.4	1964	3695	0.24	150	21	m(12c)	134		67(3+)	103(100)
Rubidium	Rb	37	85.47	1.53	39	686	0.36	58	7.6	m(8)	248		152(1+)	85(72.2), *87*(27.8)
Ruthenium	Ru	44	101.1	12.1	2334	4150	0.24	117	13	m(12h)	133		62(4+)	96(5.5), 98(1.9), 99(12.7), 100(12.6), 101(17.0), 102(31.6), 104(18.7)
Rutherfordium	Rf	104	(265.1)											
Samarium	Sm	62	150.4	7.54	1072	1791	0.20	13	1.1	m(6r)	179		96(3+)	144(3.1), *147*(15.0), *148*(11.3), *149*(13.8), 150(7.4), 152(26.7), 154(22.7)
Scandium	Sc	21	44.96	3.0	1538	2730	0.57	15.8	1.8	m(12c)	161		75(3+)	45(100)
Seaborgium	Sg	106	(271.1)											
Selenium														
(grey)	Se	34	78.96	4.80	217	685	0.32	3	10^{-4}	8a	116		198(2−), 42(6+)	74(0.9), 76(9.4), 77(7.6), 78(23.8),
(red)				4.50	d170					Se_8		116(Se_8)		80(49.6), 82(8.7)
Silicon	Si	14	28.09[(b)]	2.33	1410	3267	0.71	148	10^{-3}	d		116(Si_2H_6)	40(4+)	28(92.2), 29(4.7), 30(3.1)
Silver	Ag	47	107.9	10.5	962	2212	0.23	429	63	m(12c)	144		115(1+), 94(2+)	107(51.8), 109(48.2)
Sodium	Na	11	22.99	0.97	98	883	1.23	141	21	m(8)	186	154	102(1+)	23(100)
Strontium	Sr	38	87.62	2.6	769	1384	0.30	35.3	4.3	m(12c)	215		118(2+), 126(2+)[8]	84(0.6), 86(9.9), 87(7.0), 88(82.5)

[(b)] See addendum at the end of this table.

Element	Symbol	Z	M / g mol^{-1}	ρ / g cm^{-3}	t_m / °C	t_b / °C	c_p / J K^{-1} g^{-1}	λ / J s^{-1} m^{-1} K^{-1}	κ / MS m^{-1}	*Str.*	r_{met} / pm	r_{cov} / pm	r_{ion} / pm	Isotopes Mass number (% abundance)
Sulfur (rhombic)	S	16	32.06[b]	2.07	113	445	0.71	0.27	10^{-21}	S_8		102(H_2S_2),	184(2−), 29(6+)	32(95.0), 33(0.8), 34(4.2), 36(0.02)
(monoclinic)				1.96	119	445	0.72			S_8		104(S_8)		
Tantalum	Ta	73	180.9	16.6	3000	5458	0.14	58	8.0	m(8)	143		64(5+)	*180*(0.01), 181(99.99)
Technetium	Tc	43	(97.91)	11.5	2157	4880	0.25	51	5	m(12h)	135		65(4+), 56(7+)	
Tellurium	Te	52	127.6	6.24	450	990	0.20	3	10^{-4}	8a	143		221(2−), 97(4+), 56(6+)	120(0.1), 122(2.6), *123*(0.9), 124(4.8), 125(7.1), 127(19.0), *128*(31.7), *130*(33.8)
Terbium	Tb	65	158.9	8.25	1360	3230	0.18	11	0.9	m(12h)	176		92(3+), 76(4+)	159(100)
Thallium	Tl	81	204.4[b]	11.85	303	1457	0.13	46	5.6	m(12h)	170		150(1+), 89(3+)	203(29.5), 205(70.5)
Thorium	Th	90	232.0	11.7	1750	4788	0.12	54	7	m(12c)	180		94(4+)	*232*(100)
Thulium	Tm	69	168.9	9.32	1545	1950	0.16	17	1.4	m(12h)	172		88(3+)	169(100)
Tin (white)	Sn	50	118.7	7.30	232	2602	0.23	67	8.7	m(6t)	151		93(2+), 69(4+)	112(1.0), 114(0.7), 115(0.3), 116(14.5), 117(7.7), 118(24.2), 119(8.6), 120(32.6), 122(4.6), 124(5.8)
(grey)				5.76						d		140(d)		
Titanium	Ti	22	47.87	4.50	1660	3287	0.52	22	2.3	m(12h)	145		86(2+), 67(3+), 61(4+)	46(8.3), 47(7.4), 48(73.7), 49(5.4), 50(5.2)
Tungsten	W	74	183.8	19.3	3410	5660	0.13	174	19	m(8)	137		66(4+), 60(6+)	180(0.1), 182(26.5), 183(14.3), 184(30.7), 186(28.4)
Ununoctium	Uuo	118	[a]											
Ununpentium	Uup	115	(288.2)											
Ununseptium	Uus	117	[a]											
Ununtrium	Uut	113	(284.2)											
Uranium	U	92	238.0	19.1	1135	3818	0.12	28	3.3	m(12h)	139		89(4+), 73(6+)	*234*(0.01), *235*(0.7), *238*(99.3)
Vanadium	V	23	50.94	6.1	1910	3407	0.49	31	5.0	m(8)	131		79(2+), 64(3+), 54(5+)	*50*(0.2), 51(99.8)
Xenon	Xe	54	131.3	3.5(l)	−112	−108	0.16	0.006		(12c)		130(XeF_2)		124(0.1), 126(0.1), 128(1.9), 129(26.4), 130(4.1), 131(21.2), 132(26.9), 134(10.4), 136(8.9)

[a] The accurate molar mass has not yet been determined.

[b] See addendum at the end of this table.

Element	Symbol	Z	$\frac{M}{\text{g mol}^{-1}}$	$\frac{\rho}{\text{g cm}^{-3}}$	$\frac{t_m}{°\text{C}}$	$\frac{t_b}{°\text{C}}$	$\frac{c_p}{\text{J K}^{-1}\text{ g}^{-1}}$	$\frac{\lambda}{\text{J s}^{-1}\text{ m}^{-1}\text{ K}^{-1}}$	$\frac{\kappa}{\text{MS m}^{-1}}$	*Str.*	$\frac{r_{met}}{\text{pm}}$	$\frac{r_{cov}}{\text{pm}}$	$\frac{r_{ion}}{\text{pm}}$	Isotopes Mass number (% abundance)
Ytterbium	Yb	70	173.1	6.97	824	1194	0.15	38	3.6	m(12c)	194		87(3+)	168(0.1), 170(3.0), 171(14.3), 172(21.9), 173(16.1), 174(31.8), 176(12.7)
Yttrium	Y	39	88.91	4.47	1522	3338	0.30	17	1.7	m(12h)	178		90(3+)	89(100)
Zinc	Zn	30	65.38	7.14	420	907	0.39	116	17	m(12h)	133		74(2+)	64(48.6), 66(27.9), 67(4.1), 68(18.8), 70(0.6)
Zirconium	Zr	40	91.22	6.49	1852	4377	0.28	23	2.4	m(12h)	159		72(4+)	90(51.5), 91(11.2), 92(17.1), 94(17.4), 96(2.8)

Addendum to 9

While the distribution of isotopes in samples of most elements is essentially constant, ten elements (B, C, Cl, H, Li, N, O, Si, S and Tl) show substantial variation in their isotopic compositions, depending on the source of the element. Therefore, instead of quoting a single value for the molar mass of these elements, a range of values of molar mass is given, which corresponds to the lowest and highest values measured in natural samples. The following table gives the range of molar mass values for each of these ten elements, together with the conventional molar masses, which can be used when the source of the sample is unknown.

Atomic mass ranges, and conventional molar masses for B, C, Cl, H, Li, N, O, Si, S and Tl

Element	Symbol	Z	Molar mass range (values in g mol^{-1})	Conventional molar mass/g mol^{-1}
Boron	B	5	[10.806; 10.821]	10.81
Carbon	C	6	[12.0096; 12.0116]	12.011
Chlorine	Cl	17	[35.446; 35.457]	35.45
Hydrogen	H	1	[1.007 84; 1.008 11]	1.008
Lithium	Li	3	[6.938; 6.997]	6.94
Nitrogen	N	7	[14.006 43; 14.007 28]	14.007
Oxygen	O	8	[15.999 03; 15.999 77]	15.999
Silicon	Si	14	[28.084; 28.086]	28.085
Sulfur	S	16	[32.059; 32.076]	32.06
Thallium	Tl	81	[204.382; 204.385]	204.38

10 PROPERTIES AND APPLICATIONS OF SOME COMMON RADIOISOTOPES

Element	Isotope	Half life[a]	Mode of decay[b]	Applications
Americium	^{241}Am	432 y	α	In smoke detectors
Bismuth	^{213}Bi	46 m	β^-, α	Targeted alpha therapy
Calcium	^{47}Ca	4.5 d	β^-	Study of cellular functions and bone formation
Carbon	^{11}C	20.4 m	β^+, EC	In positron emission tomography
	^{14}C	5730 y	β^-	Determination of the age of carbon-containing artefacts
Chlorine	^{36}Cl	3.01×10^5 y	β^-, β^+, EC	Measurement of sources of chloride and the determination of the age of water
Chromium	^{51}Cr	28 d	EC	Labelling of red blood cells
Cobalt	^{57}Co	272 d	EC	Diagnosis of pernicious anaemia
	^{60}Co	5.27 y	β^-	Cancer treatment; sterilisation of biological materials
Copper	^{64}Cu	13 h	β^-, β^+, EC	Study of genetic disease affecting copper metabolism; in positron emission tomography
	^{67}Cu	2.6 d	β^-	Therapeutic use
Dysprosium	^{165}Dy	2.33 h	β^-	Treatment of arthritis
Fluorine	^{18}F	109.8 m	β^+, EC	Tracer as fluorothymidine
Gallium	^{67}Ga	78 h	EC	Medical diagnosis
	^{68}Ga	68 m	β^+, EC	In positron emission tomography
Holmium	^{166}Ho	27 h	β^-	Diagnosis and treatment of liver tumours
Indium	^{111}In	2.8 d	EC	Specialist diagnostic studies
Iodine	^{123}I	13 h	β^+, EC	Imaging to monitor thyroid function
	^{125}I	60 d	EC	Cancer brachytherapy; evaluation of filtration rate of the kidney; in radioimmuno-assays
	^{131}I	8 d	β^-	Treatment of thyroid cancer; diagnosis of abnormal liver function
Iridium	^{192}Ir	74 d	β^-	Internal radiotherapy source for cancer treatment
Iron	^{59}Fe	45 d	β^-	Studies of blood

[a] Half life: y = years; d = days; h = hours; s = seconds.
[b] Mode of decay: α = alpha particle emission; β^- = negative beta emission; β^+ = positron emission; EC = orbital electron capture; IT = isomeric transition from upper to lower isomeric state.

Element	Isotope	Half life[a]	Mode of decay[b]	Applications
Lead	^{210}Pb ^{212}Pb	22.3 y 10.6 h	β^- β^-	Dating of layers of soil and sand In targeted anticancer therapies
Lutetium	^{177}Lu	6.7 d	β^-	Therapy on small tumours
Molybdenum	^{99}Mo	66 h	β^-	Parent in the production of ^{99m}Tc
Nickel	^{63}Ni	100 y	β^-	Explosive detection; electron capture detectors for gas chromatographs
Nitrogen	^{13}N	10 m	β^+, EC	In positron emission tomography
Oxygen	^{15}O	122 s	β^+, EC	In positron emission tomography
Palladium	^{103}Pd	17 d	EC	In brachytherapy permanent implant seeds for early stage prostate cancer treatment
Phosphorus	^{32}P	14 d	β^-	Treatment of polycythaemia vera
Potassium	^{42}K ^{40}K	12.4 h 1.28×10^9 y	β^- β^-, β^+, EC	Determination of exchanged potassium in blood Potassium-argon sample dating
Rhenium	^{186}Re ^{188}Re	3.8 d 17 h	β^-, EC β^-	Pain relief in bone cancer Irradiation of coronary arteries
Samarium	^{153}Sm	47 h	β^-	Relief of pain in secondary cancers lodged in bone
Selenium	^{75}Se	120 d	EC	Study production of digestive enzymes (as selenomethionine)
Sodium	^{24}Na	15 h	β^-	Study of electrolytes in the body
Strontium	^{89}Sr	50 d	β^-	Reduction of pain of prostate and bone cancers
Technetium	^{99m}Tc	6 h	IT, β^-	Medical tracer used, for example, to locate tumours of the brain
Xenon	^{133}Xe	5 d	β^-	Ventilation studies
Ytterbium	^{169}Yb	32 d	EC	Cerebrospinal fluid studies
Yttrium	^{90}Y	64 h	β^-	Cancer brachytherapy
Zinc	^{65}Zn	244 d	β^+, EC	Used as a tracer for heavy metals in mining waste

[a] Half life: y = years; d = days; h = hours; s = seconds.
[b] Mode of decay: α = alpha particle emission; β^- = negative beta emission; β^+ = positron emission; EC = orbital electron capture; IT = isomeric transition from upper to lower isomeric state.

11 ELECTRONEGATIVITIES OF THE ELEMENTS (PAULING SCALE)

Pauling introduced the concept of electronegativity to describe the power of an atom to attract bonding electrons to itself when bonded to another atom. A number calculated from measured bond energies to represent this power of attraction is assigned to each element. The greater the difference in electronegativity of two elements, the higher the percentage ionic character of the bond between them.

1 **H** 2.20																	2 **He**
3 **Li** 0.98	4 **Be** 1.57											5 **B** 2.04	6 **C** 2.55	7 **N** 3.04	8 **O** 3.44	9 **F** 3.98	10 **Ne**
11 **Na** 0.93	12 **Mg** 1.31											13 **Al** 1.61	14 **Si** 1.90	15 **P** 2.19	16 **S** 2.58	17 **Cl** 3.16	18 **Ar**
19 **K** 0.82	20 **Ca** 1.00	21 **Sc** 1.36	22 **Ti** 1.54	23 **V** 1.63	24 **Cr** 1.66	25 **Mn** 1.55	26 **Fe** 1.83	27 **Co** 1.88	28 **Ni** 1.91	29 **Cu** 1.90	30 **Zn** 1.65	31 **Ga** 1.81	32 **Ge** 2.01	33 **As** 2.18	34 **Se** 2.55	35 **Br** 2.96	36 **Kr**
37 **Rb** 0.82	38 **Sr** 0.95	39 **Y** 1.22	40 **Zr** 1.33	41 **Nb** 1.6	42 **Mo** 2.16	43 **Tc** 2.10	44 **Ru** 2.2	45 **Rh** 2.28	46 **Pd** 2.20	47 **Ag** 1.93	48 **Cd** 1.69	49 **In** 1.78	50 **Sn** 1.96	51 **Sb** 2.05	52 **Te** 2.1	53 **I** 2.66	54 **Xe** 2.60
55 **Cs** 0.79	56 **Ba** 0.89	57–71*	72 **Hf** 1.3	73 **Ta** 1.5	74 **W** 1.7	75 **Re** 1.9	76 **Os** 2.2	77 **Ir** 2.2	78 **Pt** 2.2	79 **Au** 2.4	80 **Hg** 1.9	81 **Tl** 1.8	82 **Pb** 1.8	83 **Bi** 1.9	84 **Po** 2.0	85 **At** 2.2	86 **Rn**
87 **Fr** 0.7	88 **Ra** 0.9	89–103**															

* lanthanoid series	57 **La** 1.10	58 **Ce** 1.12	59 **Pr** 1.13	60 **Nd** 1.14	61 **Pm**	62 **Sm** 1.17	63 **Eu**	64 **Gd** 1.20	65 **Tb**	66 **Dy** 1.22	67 **Ho** 1.23	68 **Er** 1.24	69 **Tm** 1.25	70 **Yb**	71 **Lu** 1.0
** actinoid series	89 **Ac** 1.1	90 **Th** 1.3	91 **Pa** 1.5	92 **U** 1.7	93 **Np** 1.3	94 **Pu** 1.3	95 **Am**	96 **Cm**	97 **Bk**	98 **Cf**	99 **Es**	100 **Fm**	101 **Md**	102 **No**	103 **Lr**

12 ENTHALPIES OF MELTING, VAPORISATION AND ATOMISATION OF THE ELEMENTS

t_m (in °C) = melting temperature.
t_b (in °C) = boiling temperature (at 101.325 kPa).
s = sublimes; * = melts under pressure; values in parentheses are estimated.
$\Delta_{fus}H$ (in kJ mol^{-1}) = molar enthalpy of fusion or melting, i.e. from solid to liquid, per mole of species shown.
$\Delta_{vap}H$ (in kJ mol^{-1}) = molar enthalpy of vaporisation, i.e. from liquid to the vapour in equilibrium with liquid, per mole of species shown.
$\Delta_{sub}H$ (in kJ mol^{-1}) = molar enthalpy of sublimation, i.e. from solid to the vapour in equilibrium with solid, per mole of species shown.
$\Delta_{sub}H \approx \Delta_{fus}H + \Delta_{vap}H$.
$\Delta_{at}H$ (in kJ mol^{-1}) = molar enthalpy of atomisation. This is the standard enthalpy of formation of the gaseous monatomic element, at 298 K.

Note that these enthalpies are quoted per mole of atoms. This does not necessarily mean that the elements are present as isolated atoms.

For enthalpies of *compounds*, see table 16: 'Properties of elements and inorganic compounds', and table 25: 'Properties of organic compounds'.

Element	t_m / °C	t_b / °C	$\Delta_{fus}H$ at t_m / kJ mol^{-1}	$\Delta_{vap}H$ at t_b / kJ mol^{-1}	$\Delta_{vap}H$ at 298 K / kJ mol^{-1}	$\Delta_{at}H$ at 298 K / kJ mol^{-1}
Ac	1050	3200	14		406 (sub)	406
Ag	962	2212	11	258	273	285
Al	660	2467	11	291	319	330
Am	1176	(2067)	14	238	260	284
Ar	−189	−186	1.2	6.5		0
As	*817	s613	28	32	36 (sub)	302
At	302	(335)	12	33		
Au	1064	2856	12.6	324		366
B	2300	3660	50	480	565 (sub)	565
Ba	725	1640	7.7	151		180
Be	1278	2471	12	309		324
Bi	271	1560	11	179	196	207
Bk	986	(2623)				310
Br	−7	59	5.3	15	15	112
C	*3974	s3930	105		716 (sub)	717
Ca	842	1484	8.5	151	168	178
Cd	321	767	6.4	100	107	112
Ce	795	3433	5.2	314		423
Cf	900					196
Cl	−101	−34	3.2	10.2	8.8	121
Cm	1340	3110	15			386
Co	1495	2870	15.5	381		425
Cr	1857	2672	21	347	385	397
Cs	28	669	2.1	66	74	76
Cu	1085	2572	13	305	326	337
Dy	1410	2567	17	251		290
Er	1529	2900	17		317 (sub)	317
Es	(860)		9.4			133
Eu	826	1596	10.5	176		175
F	−220	−188	0.25	3.3		79
Fe	1535	2750	14	351	374	416
Fm						
Fr	(27)	(680)	2.1	64		73
Ga	30	2403	5.6	270	271	277

12 ENTHALPIES OF THE ELEMENTS *(continued)*

Element	t_m / °C	t_b / °C	$\Delta_{fus}H$ at t_m / kJ mol^{-1}	$\Delta_{vap}H$ at t_b / kJ mol^{-1}	$\Delta_{vap}H$ at 298 K / kJ mol^{-1}	$\Delta_{at}H$ at 298 K / kJ mol^{-1}
Gd	1313	3273	15.5	312		398
Ge	937	2830	32	328	334	372
H	−259	−253	0.060	0.45		218
He	*−272	−269	0.02	0.08		0
Hf	2233	4450	22	571	595	619
Hg	−39	357	2.3	59		61
Ho	1472	2694	17	251		301
I	114	184	7.9	21	23	107
In	157	2080	3.3	232		243
Ir	2450	4500	28	611		665
K	63	760	2.4	79	87	89
Kr	−157	−152	1.6	9.0		0
La	920	3470	10	402		431
Li	180	1342	2.9	148	156	159
Lr						
Lu	1663	3402	19	247		428
Md	827					
Mg	650	1110	8.4	128	138	147
Mn	1244	2095	15	220	263	281
Mo	2610	5560	36	590		658
N	−210	−196	0.36	2.8		473
Na	98	883	2.6	99	104	107
Nb	2468	4758	27	695		726
Nd	1024	3074	7.1	284		328
Ne	−249	−246	0.33	1.8		0
Ni	1455	2730	18	372	412	430
No						
Np	637	(3900)	9.5	337		465
O	−219	−183	0.22	3.4		249
Os	3033	5012	58	738		791
P	44	280	0.66	13	14	317
Pa	1572	(4000)	15	460		607
Pb	327	1740	4.8	178	190	195
Pd	1554	2970	17	393		378
Pm	1042	(2730)	13	293		
Po	254	962	13	101	109	144
Pr	935	3510	6.9	333		356
Pt	1772	3825	22	469		565
Pu	640	3230	2.8	343		364
Ra	700	1737	8	137		159
Rb	39	686	2.2	76	79	81
Re	3185	5596	34	704		770
Rh	1964	3695	27	494		557
Rn	−71	−62	2.9	16		0
Ru	2334	4150	39	592		643
S	113	445	1.7	45		277
Sb	631	1635	20	68		262
Sc	1538	2730	16	305		378

Element	t_m / °C	t_b / °C	$\Delta_{fus}H$ at t_m / kJ mol^{-1}	$\Delta_{vap}H$ at t_b / kJ mol^{-1}	$\Delta_{vap}H$ at 298 K / kJ mol^{-1}	$\Delta_{at}H$ at 298 K / kJ mol^{-1}
Se	217	685	6.7	95		227
Si	1410	3267	50	359		450
Sm	1072	1791	11	165	197	207
Sn	232	2602	7.1	296		301
Sr	769	1384	7.4	139	154	164
Ta	3000	5458	37	733		782
Tb	1360	3230	16	293		389
Tc	2157	4880	23	577		678
Te	450	990	17	114		197
Th	1750	4788	14	544		602
Ti	1660	3287	15.5	429		473
Tl	303	1457	4.3	166	175	182
Tm	1545	1950	18	213		232
U	1135	3818	15	423		533
V	1910	3407	18	459		514
W	3410	5660	35	824		849
Xe	−112	−108	2.3	12.6		0
Y	1522	3338	17	367		421
Yb	824	1194	7.7	159		152
Zn	420	907	7.4	115	123	130
Zr	1852	4377	19	582		609

13 FIRST IONISATION ENTHALPIES (E_{i1}) OF THE ELEMENTS

The first ionisation energy of an element is the change in internal energy at 0 K, $\Delta U_{0\,K}$, which accompanies the ionisation of one mole of gaseous atoms to form one mole of gaseous univalent cations. It is therefore defined as being the energy change for the reaction

$$X_{(g)} \rightarrow X^+_{(g)} + e^-_{(g)}$$

The enthalpy change at 298 K (25 °C) for this process, $\Delta H_{298\,K}$, is related to $\Delta U_{0\,K}$ by the following:

$$\Delta H_{298\,K} = \Delta U_{0\,K} + \int_{0\,K}^{298\,K} [C_p(X^+) + C_p(e^-) - C_p(X)]\,dT$$

If $X^+_{(g)}$, $e^-_{(g)}$ and $X_{(g)}$ are assumed to be perfect gases, their heat capacities, C_p, may be taken as zero at 0 K, and $(5/2)R$ at other temperatures. Therefore

$$\Delta H_{298\,K} = \Delta U_{0\,K} + \int_{0\,K}^{298\,K} \frac{5}{2} R\,dT$$
$$= \Delta U_{0\,K} + 6.2\,\text{kJ mol}^{-1}$$

Values of ionisation energies are also commonly reported in electronvolts, eV (a non-SI unit), in which case they are called ionisation potentials, *I*.

In the periodic table on the next page, first ionisation enthalpies at 298 K ($\Delta H_{298\,K}$/kJ mol^{-1}) are given directly below the element symbol, while first ionisation potentials at 0 K (*I*/eV) are given in parentheses.

$$1\ \text{eV} = 96.485\ 308\ 91\ \text{kJ mol}^{-1}$$

1 **H** 1318 (13.60)																	2 **He** 2378 (24.59)
3 **Li** 526 (5.39)	4 **Be** 906 (9.32)											5 **B** 807 (8.30)	6 **C** 1092 (11.26)	7 **N** 1408 (14.53)	8 **O** 1320 (13.62)	9 **F** 1687 (17.42)	10 **Ne** 2087 (21.56)
11 **Na** 502 (5.14)	12 **Mg** 744 (7.65)											13 **Al** 584 (5.99)	14 **Si** 793 (8.15)	15 **P** 1018 (10.49)	16 **S** 1006 (10.36)	17 **Cl** 1257 (12.97)	18 **Ar** 1527 (15.76)
19 **K** 425 (4.34)	20 **Ca** 596 (6.11)	21 **Sc** 648 (6.56)	22 **Ti** 665 (6.83)	23 **V** 657 (6.75)	24 **Cr** 659 (6.77)	25 **Mn** 723 (7.43)	26 **Fe** 768 (7.90)	27 **Co** 766 (7.88)	28 **Ni** 743 (7.64)	29 **Cu** 751 (7.73)	30 **Zn** 912 (9.39)	31 **Ga** 585 (6.00)	32 **Ge** 768 (7.90)	33 **As** 950 (9.79)	34 **Se** 947 (9.75)	35 **Br** 1146 (11.81)	36 **Kr** 1357 (14.00)
37 **Rb** 409 (4.18)	38 **Sr** 555 (5.69)	39 **Y** 606 (6.22)	40 **Zr** 646 (6.63)	41 **Nb** 658 (6.76)	42 **Mo** 690 (7.09)	43 **Tc** 708 (7.28)	44 **Ru** 716 (7.36)	45 **Rh** 726 (7.46)	46 **Pd** 810 (8.34)	47 **Ag** 737 (7.58)	48 **Cd** 874 (8.99)	49 **In** 564 (5.79)	50 **Sn** 715 (7.34)	51 **Sb** 837 (8.61)	52 **Te** 875 (9.01)	53 **I** 1014 (10.45)	54 **Xe** 1176 (12.13)
55 **Cs** 382 (3.89)	56 **Ba** 509 (5.21)	57–71 *	72 **Hf** 665 (6.83)	73 **Ta** 734 (7.55)	74 **W** 775 (7.86)	75 **Re** 762 (7.83)	76 **Os** 820 (8.44)	77 **Ir** 871 (8.97)	78 **Pt** 870 (8.96)	79 **Au** 896 (9.23)	80 **Hg** 1013 (10.44)	81 **Tl** 595 (6.11)	82 **Pb** 722 (7.42)	83 **Bi** 709 (7.29)	84 **Po** 818 (8.41)	85 **At**	86 **Rn** 1043 (10.75)
87 **Fr** 399 (4.07)	88 **Ra** 515 (5.28)	89–103 **															

* lanthanoid series	57 **La** 544 (5.58)	58 **Ce** 540 (5.54)	59 **Pr** 534 (5.47)	60 **Nd** 539 (5.53)	61 **Pm** 545 (5.58)	62 **Sm** 551 (5.64)	63 **Eu** 553 (5.67)	64 **Gd** 599 (6.15)	65 **Tb** 572 (5.86)	66 **Dy** 579 (5.94)	67 **Ho** 587 (6.02)	68 **Er** 595 (6.11)	69 **Tm** 603 (6.18)	70 **Yb** 609 (6.25)	71 **Lu** 530 (5.43)
** actinoid series	89 **Ac** 505 (5.17)	90 **Th** 593 (6.31)	91 **Pa** 574 (5.89)	92 **U** 604 (6.19)	93 **Np** 611 (6.27)	94 **Pu** 591 (6.03)	95 **Am** 584 (5.99)	96 **Cm** 587 (6.00)	97 **Bk** 607 (6.19)	98 **Cf** 614 (6.28)	99 **Es** 626 (6.42)	100 **Fm** 633 (6.5)	101 **Md** 641 (6.58)	102 **No** 648 (6.65)	103 **Lr** 479 (4.9)

14 SUCCESSIVE IONISATION ENTHALPIES (E_{in}) OF THE ELEMENTS

The first ionisation enthalpy is the enthalpy change for the process

$$X_{(g)} \rightarrow X^{+}_{(g)} + e^{-}_{(g)} \qquad \Delta H = E_{i1}$$

The second ionisation enthalpy is the enthalpy change for the process

$$X^{+}_{(g)} \rightarrow X^{2+}_{(g)} + e^{-}_{(g)} \qquad \Delta H = E_{i2}$$

Each successive ionisation enthalpy is defined in a similar way.

The values in this table refer to 298 K and are given in megajoules per mole of electrons released (MJ mol^{-1}). This is a convenient unit to show the significant figures to which the values are known.

Z	Element	E_{i1}	E_{i2}	E_{i3}	E_{i4}	E_{i5}	E_{i6}	E_{i7}	E_{i8}	E_{i9}	E_{i10}	E_{i11}	E_{i12}	E_{i13}	E_{i14}	E_{i15}	E_{i16}	E_{i17}	E_{i18}	E_{i19}	E_{i20}
1	H	1.318																			
2	He	2.379	5.257																		
3	Li	0.526	7.305	11.822																	
4	Be	0.906	1.763	14.855	21.013																
5	B	0.807	2.433	3.666	25.033	32.834															
6	C	1.093	2.359	4.627	6.229	37.838	47.285														
7	N	1.407	2.862	4.585	7.482	9.452	53.274	64.368													
8	O	1.320	3.395	5.307	7.476	10.996	13.333	71.343	84.086												
9	F	1.687	3.381	6.057	8.414	11.029	15.171	17.874	92.047	106.443											
10	Ne	2.087	3.959	6.128	9.376	12.184	15.245	20.006	23.076	115.389	131.442										
11	Na	0.502	4.569	6.919	9.550	13.356	16.616	20.121	25.497	28.941	141.373	159.086									
12	Mg	0.744	1.457	7.739	10.547	13.636	18.001	21.710	25.663	31.650	35.469	170.003	189.379								
13	Al	0.584	1.823	2.751	11.584	14.837	18.384	23.302	27.466	31.905	38.464	42.661	201.283	222.327							
14	Si	0.793	1.583	3.238	4.362	16.098	19.791	23.793	29.259	33.884	38.740	45.941	50.519	235.218	257.932						
15	P	1.018	1.909	2.918	4.963	6.280	21.275	25.404	29.861	35.864	40.966	46.280	54.080	59.043	271.813	296.208					
16	S	1.006	2.257	3.367	4.570	7.019	8.502	27.113	31.676	36.585	43.146	48.712	54.489	62.882	68.238	311.074	337.15				
17	Cl	1.257	2.303	3.828	5.164	6.548	9.368	11.025	33.612	38.607	43.968	51.074	57.096	63.370	72.348	78.104	353.01	380.77			
18	Ar	1.527	2.672	3.937	5.777	7.245	8.787	12.002	13.848	40.767	46.194	52.009	59.660	66.207	72.926	82.481	88.6	397.62	427.08		
19	K	0.425	3.058	4.418	5.883	7.982	9.66	11.349	14.948	16.971	48.583	54.439	60.707	68.902	75.956	83.158	93.4	99.8	444.92	476.08	
20	Ca	0.596	1.152	4.918	6.480	8.150	10.502	12.33	14.213	18.198	20.391	57.056	63.341	70.061	78.801	86.376	94.0	104.9	111.6	494.91	527.78

1 electronvolt (eV) = 96.485 308 91 kJ mol^{-1} = 0.096 485 308 91 MJ mol^{-1}.

14 SUCCESSIVE IONISATION ENTHALPIES (E_{in}) OF THE ELEMENTS *(continued)*

Z	Element	E_{i1}	E_{i2}	E_{i3}	E_{i4}	E_{i5}	E_{i6}	E_{i7}
21	Sc	0.637	1.241	2.395	7.095	8.850	10.73	13.32
22	Ti	0.664	1.316	2.659	4.181	9.580	11.523	13.59
23	V	0.656	1.420	2.834	4.513	6.300	12.368	14.496
24	Cr	0.659	1.598	2.993	4.74	6.69	8.744	15.55
25	Mn	0.724	1.515	3.255	4.95	6.99	9.2	11.514
26	Fe	0.766	1.567	2.964	5.29	7.24	9.6	12.1
27	Co	0.765	1.652	3.238	4.96	7.68	9.8	12.5
28	Ni	0.743	1.759	3.400	5.30	7.29	10.4	12.8
29	Cu	0.752	1.964	3.560	5.33	7.72	9.9	13.4
30	Zn	0.913	1.740	3.839	5.74	7.98	10.4	12.9
31	Ga	0.585	1.985	2.969	6.2			
32	Ge	0.768	1.544	3.308	4.4	9.03		
33	As	0.953	1.804	2.742	4.843	6.049	12.32	
34	Se	0.947	2.051	2.980	4.150	6.60	7.889	15.00
35	Br	1.146	2.11	3.5	4.57	5.77	8.56	9.94
36	Kr	1.357	2.374	3.571	5.07	6.25	7.58	10.72
37	Rb	0.409	2.638	3.9	5.08	6.86	8.15	9.58
38	Sr	0.556	1.071	4.21	5.5	6.92	8.77	10.2
39	Y	0.606	1.187	1.986	5.97	7.44	8.98	11.20
40	Zr	0.666	1.273	2.224	3.319	7.87		
41	Nb	0.670	1.388	2.422	3.70	4.884	9.91	12.1
42	Mo	0.691	1.564	2.627	4.48	5.91	6.6	12.24
43	Tc	0.708	1.478	2.856				
44	Ru	0.717	1.623	2.753				
45	Rh	0.726	1.751	3.003				

Z	Element	E_{i1}	E_{i2}	E_{i3}	E_{i4}	E_{i5}	E_{i6}	E_{i7}
46	Pd	0.811	1.881	3.183				
47	Ag	0.737	2.080	3.367				
48	Cd	0.874	1.638	3.622				
49	In	0.565	1.827	2.711	5.2			
50	Sn	0.715	1.418	2.949	3.937	6.980		
51	Sb	0.840	1.601	2.45	4.27	5.4	10.4	
52	Te	0.876	1.80	2.704	3.616	5.675	6.83	13.2
53	I	1.015	1.852	3.2				
54	Xe	1.177	2.053	3.10				
55	Cs	0.382	2.43					
56	Ba	0.509	0.972					
57	La	0.544	1.073	1.857	4.826	5.95		
58	Ce	0.541	1.053	1.955	3.553	6.33	7.5	
59	Pr	0.533	1.024	2.092	3.767			
60	Nd	0.539	1.042	2.14	3.91			
61	Pm	0.542	1.056	2.16	3.97			
62	Sm	0.551	1.074	2.26	4.00			
63	Eu	0.553	1.091	2.41	4.12			
64	Gd	0.600	1.17	2.00	4.25			
65	Tb	0.572	1.118	2.12	3.81			
66	Dy	0.579	1.132	2.13	4.01			
67	Ho	0.587	1.145	2.21	4.11			
68	Er	0.596	1.157	2.20	4.12			
69	Tm	0.603	1.169	2.29	4.13			
70	Yb	0.610	1.180	2.421	4.21			

1 electronvolt (eV) = 96.485 308 91 kJ mol^{-1} = 0.096 485 308 91 MJ mol^{-1}.

14 SUCCESSIVE IONISATION ENTHALPIES (E_{in}) OF THE ELEMENTS *(continued)*

Z	Element	E_{i1}	E_{i2}	E_{i3}	E_{i4}	E_{i5}	E_{i6}	E_{i7}
71	Lu	0.530	1.35	2.028	4.37			
72	Hf	0.660	1.44	2.25	3.222			
73	Ta	0.767						
74	W	0.776						
75	Re	0.766						
76	Os	0.85						
77	Ir	0.88						
78	Pt	0.87	1.797					
79	Au	0.896	1.98					
80	Hg	1.013	1.816	3.31				
81	Tl	0.596	1.977	2.884				
82	Pb	0.722	1.457	3.088	4.089			
83	Bi	0.710	1.616	2.472	4.38			
84	Po	0.818						
85	At							
86	Rn	1.043						
87	Fr							
88	Ra	0.516	0.985					
89	Ac	0.505	1.17					
90	Th	0.593	1.12	1.94	2.79			
91	Pa	0.574						
92	U	0.604						
93	Np	0.611						
94	Pu	0.591						
95	Am	0.584						

1 electronvolt (eV) = 96.485 308 91 kJ mol^{-1} = 0.096 485 308 91 MJ mol^{-1}.

15 ELECTRON AFFINITIES (E_{ea}) OF THE ELEMENTS

The electron affinity of an atom X can be defined in terms of either internal energy or enthalpy. Both definitions are given below, with the tabulated data given in terms of enthalpy.

(i) Internal energy

The electron affinity of an atom X is the change in internal energy at 0 K, $\Delta U_{0\,K}$, which accompanies the loss of one mole of electrons from one mole of gaseous negative ions of the atom, to form one mole of gaseous atoms. It is therefore defined as being the energy change for the reaction

$$X^-{}_{(g)} \rightarrow X_{(g)} + e^-{}_{(g)}$$

(ii) Enthalpy

The enthalpy change at 298 K (25 °C) accompanying the formation of one mole of gaseous atoms X from one mole of gaseous negative ions X^- is related to the $\Delta U_{0\,K}$ value by the following:

$$\Delta H_{298\,K} = \Delta U_{0\,K} + \int_{0\,K}^{298\,K} [C_p(X) + C_p(e^-) - C_p(X^-)]\,dT$$

If it is assumed that C_p for the gaseous species X^-, X and e^- is equal to zero at 0 K, and $(5/2)R$ at other temperatures, then at 298 K

$$\Delta H_{298\,K} = \Delta U_{0\,K} + \frac{5}{2}RT = \Delta U_{0\,K} + 6.2 \text{ kJ mol}^{-1}$$

The electron affinity is tabulated as $\Delta H_{298\,K}$ for the process

$$X^-{}_{(g)} \rightarrow X_{(g)} + e^-{}_{(g)}$$

On the next page electron affinities at 298 K ($\Delta H_{298\,K}$/kJ mol^{-1}) are given directly below the element symbol, while electron affinities measured in electronvolts ($\Delta H_{298\,K}$/eV) are given in parentheses.

1 eV = 96.485 308 91 kJ mol^{-1}
ns = not stable

1 **H** 79 (0.75)																	2 **He** ns
3 **Li** 66 (0.62)	4 **Be** ns											5 **B** 33 (0.28)	6 **C** 128 (1.26)	7 **N** ns	8 **O** 147 (1.46)	9 **F** 334 (3.40)	10 **Ne** ns
11 **Na** 59 (0.55)	12 **Mg** ns											13 **Al** 48 (0.43)	14 **Si** 140 (1.39)	15 **P** 78 (0.75)	16 **S** 206 (2.08)	17 **Cl** 355 (3.61)	18 **Ar** ns
19 **K** 54 (0.50)	20 **Ca** 8 (0.02)	21 **Sc** 24 (0.19)	22 **Ti** 14 (0.08)	23 **V** 57 (0.53)	24 **Cr** 70 (0.67)	25 **Mn** ns	26 **Fe** 21 (0.15)	27 **Co** 70 (0.66)	28 **Ni** 118 (1.16)	29 **Cu** 125 (1.24)	30 **Zn** ns	31 **Ga** 47 (0.43)	32 **Ge** 125 (1.23)	33 **As** 84 (0.80)	34 **Se** 201 (2.02)	35 **Br** 331 (3.36)	36 **Kr** ns
37 **Rb** 53 (0.49)	38 **Sr** 11 (0.05)	39 **Y** 36 (0.31)	40 **Zr** 47 (0.43)	41 **Nb** 94 (0.92)	42 **Mo** 78 (0.75)	43 **Tc** 59 (0.55)	44 **Ru** 107 (1.05)	45 **Rh** 116 (1.14)	46 **Pd** 60 (0.56)	47 **Ag** 132 (1.30)	48 **Cd** ns	49 **In** 35 (0.3)	50 **Sn** 113 (1.11)	51 **Sb** 107 (1.05)	52 **Te** 196 (1.97)	53 **I** 301 (3.06)	54 **Xe** ns
55 **Cs** 52 (0.47)	56 **Ba** 20 (0.14)	57–71 *	72 **Hf** 7.0 (0.01)	73 **Ta** 37 (0.32)	74 W 85 (0.82)	75 **Re** 20 (0.15)	76 **Os** 112 (1.1)	77 **Ir** 157 (1.56)	78 **Pt** 211 (2.13)	79 **Au** 229 (2.31)	80 **Hg** ns	81 **Tl** 42 (0.38)	82 **Pb** 41 (0.36)	83 **Bi** 97 (0.94)	84 **Po** 189 (1.9)	85 **At** 276 (2.8)	86 **Rn** ns
87 **Fr** 53 (0.49)	88 **Ra** 16 (0.10)	89 **Ac** 40 (0.35)															

* lanthanoid series	57 **La** 51 (0.47)	58 **Ce** 69 (0.65)	59 **Pr** 99 (0.96)	60 **Nd** 191 (>1.916)	61 **Pm**	62 **Sm**	63 **Eu** 89 (0.86)	64 **Gd**	65 **Tb** 118 (>1.165)	66 **Dy** >0 (>0)	67 **Ho**	68 **Er**	69 **Tm** 105 (1.03)	70 **Yb** 4 (–0.020)	71 **Lu** 39 (0.34)

16 PROPERTIES OF ELEMENTS AND INORGANIC COMPOUNDS

Note that all elements and compounds listed in the following table should be assumed to be hazardous. Appropriate safety information is available in the relevant MSDS, which may be obtained from www.sigmaaldrich.com/safety-center.html or other online sources.

The following table contains names, formulae, hazards, structures of binary compounds, physical properties, thermochemical data, and dipole moments of elements and selected inorganic compounds.

Formula = empirical or molecular formula.

Structure The structure of solid binary compounds is indicated by the number on the right-hand side of the formula column. Many of these structures are illustrated in table 17: 'Some crystal forms'.

0 = molecular
- 0(d) = dimer
- 0(t) = trimer

1 = sodium chloride
2 = caesium chloride
3 = zinc blende
4 = wurtzite
5 = nickel arsenide
6 = fluorite or antifluorite
7 = rutile
8 = layer structures
- 8i = cadmium chloride or anticadmium chloride
- 8ii = cadmium iodide
- 8iii = chromium(III) chloride
- 8iv = bismuth iodide
- 8v = other layer structures
- 8a = chain

9 = structure other than 0 to 8a
– = no structural details known
* = distorted structure
p = polymer
V = solid structure unknown, but the molecular structure of the vapour is shown by the formula. The shapes of many of these molecules are given in table 18: 'Shapes of some molecules and ions'.
- V(d) = dimer

M (in g mol^{-1}) = molar mass for the formula shown.

col. = colour:
b = blue; bk = black; bn = brown; g = green; gy = grey; nil = colourless; o = orange; p = pink; r = red; s = silver; v = violet; w = white; y = yellow.

ρ (in g cm^{-3}) = density at 298 K (l = liquid).

t_m (in °C) = melting temperature.

t_b (in °C) = boiling temperature (at 101.325 kPa).

d = decomposes; s = sublimes; * = melts under pressure; values in parentheses are estimated.

sol. = solubility. This is reported in all cases as mass in g of anhydrous substance per 100 g water required to prepare a saturated solution at 298 K; i = practically insoluble; s = soluble; vs = very soluble; ∞ = completely miscible; hyd = hydrolysed by water. A number in parentheses next to the solubility indicates that a hydrated form of the solid is in equilibrium with the saturated solution, and gives the number of water molecules in the hydrate. Where the solute is a gas, the solubility is at a total pressure, for the gas plus H_2O, of 101.325 kPa.

All thermochemical data are reported for the standard state pressure of 10^5 Pa and a temperature of 298 K.

s = solid; l = liquid; g = gas; aq = infinite dilution in aqueous solution.

$\Delta_f H^{\ominus}$ (in kJ mol^{-1}) = the standard enthalpy of formation for 1 mole of the substance, in the state specified, under standard state conditions, from its elements in their standard reference states (stable forms).

$\Delta_f G^\ominus$ (in kJ mol^{-1}) = the standard Gibbs energy of formation for 1 mole of the substance, in the state specified, under standard state conditions, from its elements in their standard reference states (stable forms).

$S^\ominus$ (in J K^{-1} mol^{-1}) = the standard entropy of the substance in the state specified.

$C_p^\ominus$ (in J K^{-1} mol^{-1}) = the standard molar heat capacity at constant pressure of the substance in the state specified.

Thermochemical properties for certain ionic species, principally of the elements, are also included. Ions are listed with the relevant element or related acid. In the case of aqueous ions, thermochemical values are given at infinite dilution relative to zero for the hydrogen ion; i.e. $\Delta_f H^\ominus$, $\Delta_f G^\ominus$, $S^\ominus$ and $C_p^\ominus$ are given values of zero for H^+(aq). For example the values of $\Delta_f H^\ominus$ given for Na^+(aq) and Cl^-(aq) are actually for

$$Na(s) + H^+(aq) \longrightarrow Na^+(aq) + \tfrac{1}{2}H_2(g)$$

and

$$\tfrac{1}{2}Cl_2(g) + \tfrac{1}{2}H_2(g) + aq \longrightarrow Cl^-(aq) + H^+(aq)$$

respectively.

In the case of gaseous ions it is assumed that, for the gaseous electron, $\Delta_f H^\ominus$ is zero and $S^\ominus$ is 21.0 J K^{-1} mol^{-1}.

$\Delta_{fus} H$ (in kJ mol^{-1}) = the molar enthalpy of fusion or melting, i.e. from solid to liquid, per mole of species shown.

$\Delta_{sub} H$ (in kJ mol^{-1}) = the molar enthalpy of sublimation, i.e. from solid to the vapour in equilibrium with solid, per mole of species shown.

---

p = electric dipole moment for molecules in the gas phase. Values are given in 10^{-30} C m.

1 debye (D) = 3.336×10^{-30} C m.

The permanent dipole moment of an isolated molecule is defined as

$$p = (\sum_i e_i r_i)$$

where e_i = charges on all nuclei and electrons

r_i = position vectors referred to any origin.

The dipole moments listed are for molecules of the dominant isotopic species at 0 K. The average moment determined from bulk dielectric properties at ordinary temperatures is not likely to differ by more than 1 per cent from the values given.

Substance	*Formula/Structure*	M / g mol^{-1}	*col.*	ρ / g cm^{-3}	t_m / °C	t_b / °C	*sol.* / g/100 g	*State*	$\Delta_f H^\circ$ / kJ mol^{-1}	$\Delta_f G^\circ$ / kJ mol^{-1}	S° / J K^{-1} mol^{-1}	C_p° / J K^{-1} mol^{-1}	$\Delta_{fus} H$ / kJ mol^{-1}	$\Delta_{sub} H$ / kJ mol^{-1}	p / 10^{-30} C m
Aluminium	Al	26.98	s	2.7	660	2467	i	s	0	0	28	24	11	330	
	Al							g	330	290	165	21			
	Al^{3+}							g	5483						
	Al^{3+}							aq	−538	−492	−325				
ammonium sulfate	$AlNH_4(SO_4)_2$	237.1	w	2.4				s	−2352	−2038	216	226			
	$\cdot 12H_2O$	453.3	w	1.6	93	d120	9	s	−5942	−4937	697	683			
bromide	$AlBr_3$ 0(d)	266.7	w	3.2	97	255	hyd	s	−511	−489	180	101	11	101	
	$\cdot 6H_2O$	374.8	w-y	2.5	93	d135	hyd								
	$\cdot 15H_2O$	536.9	w		−7	d7	s								
carbide	Al_4C_3 9	144.0	y-g	2.4	2100	d2220	hyd	s	−209	−196	89	117			
chloride	$AlCl_3$ 8iii	133.3	w	2.4	*193	s180	hyd	s	−704	−629	111	92	36	121	6.6
	$\cdot 6H_2O$	241.4	w	2.4	d100		45	s	−2692	−2261	318	296			
fluoride	AlF_3 9	84.0	w	2.9		s1291	0.6	s	−1510	−1431	66	75	98	299	
hydride	AlH_3 p	30.0	w		d150		hyd	s	−46						
hydroxide	$Al(OH)_3$	78.0	w	2.4	d300		i	s	−1276						
tetrahydroxidoaluminate	$Al(OH)_4^-$	95.0						aq	−1502	−1305	103				
iodide	AlI_3 0(d)	407.7	w	4.0	191	382	hyd	s	−314	−301	159	99	17	106	
nitrate	$Al(NO_3)_3$	213.0						s	−3757						
	$\cdot 9H_2O$	375.1	w	1.7	73	d150	64								
nitride	AlN 4	41.0	w	3.3	*2200	s2000	hyd	s	−318	−287	20	30		753	
oxide	Al_2O_3 9	102.0	w	4.0	2054	2980	i	s	−1676	−1582	51	79	109		
phosphate	$AlPO_4$	122.0	w	2.6	>1500		i	s	−1734	−1618	91	93			
phosphide	AlP 3	58.0	gy	2.4	2550		hyd	s	−167	−160	47				
potassium sulfate	$AlK(SO_4)_2$	258.2						s	−2470	−2240	205	193			
	$\cdot 12H_2O$	474.4	w	1.8	d93		7.8	s	−6062	−5141	687	651	28		
sulfate	$Al_2(SO_4)_3$	342.1	w	2.7	d770		39($16H_2O$)	s	−3441	−3100	239	259			
	$\cdot 18H_2O$	666.4	w	1.7	d87			s	−8879						
sulfide	Al_2S_3 9, 9	150.2	y	2.0	1100		hyd	s	−724						
Ammonia	NH_3 0	17.0	nil	0.8(l)	−78	−33	46	g	−46	−16	193	35	5.7	26	4.9
	NH_3							aq	−80	−27	111				

Substance	*Formula/Structure*	M / g mol^{-1}	*col.*	ρ / g cm^{-3}	t_m / °C	t_b / °C	*sol.* / g/100 g	*State*	$\Delta_f H^{\circ}$ / kJ mol^{-1}	$\Delta_f G^{\circ}$ / kJ mol^{-1}	S° / J K^{-1} mol^{-1}	C_p° / J K^{-1} mol^{-1}	$\Delta_{fus}H$ / kJ mol^{-1}	$\Delta_{sub}H$ / kJ mol^{-1}	p / 10^{-30} C m
Ammonium	NH_4^+	18.0						aq	−133	−79	111	80			
acetate	$NH_4C_2H_3O_2$	77.1	w	1.2	114	d	vs	s	−616						
aluminium sulfate	See Aluminium, page 37														
bromide	NH_4Br	97.9	w	2.4		s452	78	s	−271	−175	113	96			
carbamate	$NH_2CO_2NH_4$	78.1	w		d59		vs	s	−645	−448	134				
carbonate	$NH_4HCO_3 \cdot NH_2CO_2NH_4$ [(a)]	157.1	w		d58		$25^{15°C}$								
hydrogen	NH_4HCO_3	79.1	w	1.6	d36		25	s	−849	−666	121				
chlorate, per	NH_4ClO_4	117.5	w	2.0	d240			s	−295	−89	186				
chloride	NH_4Cl	53.5	w	1.5		s340	39	s	−314	−203	95	84			
dichromate	$(NH_4)_2Cr_2O_7$	252.1	o	2.2	d180		40	s	−1807						
fluoride	NH_4F	37.0	w	1.0	d		83	s	−464	−349	72	65			
hydrogen	NH_4HF_2	57.0	w	1.5	125		78	s	−803	−651	116	107			
iodide	NH_4I	144.9	w	2.5	d551		181	s	−201	−113	117	82	21		
iron(II) sulfate	$(NH_4)_2Fe(SO_4)_2$	284.0													
hexahydrate	$\cdot 6H_2O$	392.1	g	1.9	d100		25								
iron(III) sulfate	$NH_4Fe(SO_4)_2$	266.0	w	2.5	d420										
dodecahydrate	$\cdot 12H_2O$	482.2	v	1.7	40	d230	48								
molybdate	$(NH_4)_6Mo_7O_{24} \cdot 4H_2O$	1235.9	w	2.5	d		s								
nitrate	NH_4NO_3	80.0	w	1.7	169	d210	208	s	−366	−184	151	139	6.4		
oxalate	$(NH_4)_2C_2O_4$	124.1						s	−1123						
	$\cdot H_2O$	142.1	w	1.5	d70		5	s	−1425						
phosphate	$(NH_4)_3PO_4$	149.1						s	−1672						
	$\cdot 3H_2O$	203.1	w				23	s	−2556						
hydrogen	$(NH_4)_2HPO_4$	132.1	w	1.6	d155		70	s	−1567			188			
dihydrogen	$NH_4H_2PO_4$	115.0	w	1.8	190		40	s	−1445	−1210	152	142			
sulfate	$(NH_4)_2SO_4$	132.1	w	1.8	d280		76	s	−1181	−902	220	187			
hydrogen	NH_4HSO_4	115.1	w	1.8	147	d350	100	s	−1027						
peroxydi-	$(NH_4)_2S_2O_8$	228.2	w	2.0	d120		84	s	−1648						

[(a)] $(NH_4HCO_3 \cdot NH_2CO_2NH_4)$ – approximate formula of the precipitated carbonate.

Substance	*Formula/Structure*		M / g mol^{-1}	*col.*	ρ / g cm^{-3}	t_m / °C	t_b / °C	*sol.* / g/100 g	*State*	$\Delta_f H^\circ$ / kJ mol^{-1}	$\Delta_f G^\circ$ / kJ mol^{-1}	S° / J K^{-1} mol^{-1}	C_p° / J K^{-1} mol^{-1}	$\Delta_{fus}H$ / kJ mol^{-1}	$\Delta_{sub}H$ / kJ mol^{-1}	p / 10^{-30} C m
Ammonium (cont.)																
sulfide	$(NH_4)_2S$		68.1	y		d		vs	aq	−232	−73	212				
hydrogen	NH_4HS		51.1	w	1.2	d25		vs	s	−157	−51	97				
penta-	$(NH_4)_2S_5$		196.4	y		d115		vs	s	−274						
tartrate	$(NH_4)_2C_4H_4O_6$		184.1	w	1.6	d		$58^{15°C}$								
thiocyanate	NH_4SCN		76.1	w	1.3	149	d170	190	s	−79						
vanadate, meta-	NH_4VO_3		117.0	w	2.3	d200		0.6	s	−1053	−888	141	129			
Antimony	Sb		121.8	s	6.7	631	1635	i	s	0	0	46	25	20	123	
	Sb								g	262	222	180	21			
bromide, tri-	$SbBr_3$	0	361.5	w	4.3	97	288	hyd	s	−259	−239	207	108	15	65	
chloride, tri-	$SbCl_3$	0	228.1	w	3.1	73	220	hyd	s	−382	−324	184	108	13	68	13.1
chloride, penta-	$SbCl_5$	0	299.0	w	2.3	4	d140	hyd	l	−440	−350	301		10	56	0
fluoride, tri-	SbF_3	0	178.8	w	4.4	292	376	492	s	−916				21		
fluoride, penta-	SbF_5	–	216.7	w	3.0	8	141	hyd								
hydride, tri-(stibine)	SbH_3	0	124.8	nil	2.3(l)	−88	−17	s	g	145	148	233	41			0.4
iodide, tri-	SbI_3	8iv	502.5	r	4.9	171	401	hyd	s	−100	−99	215	98	18		
(di-)oxide, tri-	Sb_2O_3	8a, 0(d)	291.5	w	5.2	655	1425	0.0008	s	−720	−634	110	101	54		
(di-)oxide, tetra-	Sb_2O_4	9	307.5	w	5.8	d930		4×10^{-6}	s	−908	−796	127	115			
(di-)oxide, penta-	Sb_2O_5	9	323.5	y	3.8	d380		i	s	−972	−829	125	118			
(di-)sulfide, tri-	Sb_2S_3 (black)	8a	339.7	bk	4.6	550	(1150)	i	s(black)	−175	−174	182	120	63		
	Sb_2S_3 (orange)	8a	339.7	o	4.1	550	(1150)	i	s(orange)	−147						
(di-)sulfide, penta-	Sb_2S_5		403.8	o	4.1	d75										
Arsenic	As (grey)		74.92	gy	5.7	*817	s613	i	s(α, grey)	0	0	35	25	28	36	
	As (yellow)		74.92	y	2.0	d358		i	s(γ, yellow)	15						
	As								s(β)	4						
	As								g	302	261	174	21			
	As_2		149.8						g	222	172	239	35			
	As_4		299.7						g	144	92	314				
acid, ortho-	H_3AsO_4		141.9													
	$\cdot\frac{1}{2}H_2O$		151.0	w	2.2	35	d160	vs	s	−906						

Substance	*Formula/Structure*		M g mol^{-1}	*col.*	ρ g cm^{-3}	t_m °C	t_b °C	*sol.* g/100 g	*State*	$\Delta_f H^\circ$ kJ mol^{-1}	$\Delta_f G^\circ$ kJ mol^{-1}	S° J K^{-1} mol^{-1}	C_p° J K^{-1} mol^{-1}	$\Delta_{fus}H$ kJ mol^{-1}	$\Delta_{sub}H$ kJ mol^{-1}	p 10^{-30} C m
Arsenic (cont.)																
bromide, tri-	$AsBr_3$	0	314.6	y	3.4	31	221	hyd	g	−130	−159	364	79	12	67	
chloride, tri-	$AsCl_3$	0	181.3	nil	2.2	−16	130	hyd	l	−305	−259	216	133	10	54	5.3
fluoride, tri-	AsF_3	0	131.9	nil	2.7	−6	58	hyd	g	−786	−771	289	66	10	46	8.6
fluoride, penta-	AsF_5	0	169.9	nil		−80	−53	hyd	g	−1237	−1170	317	98	11		
hydride, tri-(arsine)	AsH_3	0	78.0	nil	1.7(l)	−116	−63	0.07	g	66	69	223	38	2		0.7
iodide, tri-	AsI_3	8iv	455.6	r	4.4	146	403	6	s	−58	−59	213	106	9.2		3.2
(di-)oxide, tri-	As_2O_3	8v, 0(d)	197.8	w	3.7	313	465	2	s	−657	−576	107	96	18	52	
(di-)oxide, penta-	As_2O_5	V(d)	229.8	w	4.3	d315		68	s	−925	−782	105	116			
(di-)sulfide, tri-	As_2S_3	8v	246.0	y	3.4	312	707	5×10^{-5}	s	−169	−169	164	116	29		
(di-)sulfide, penta-	As_2S_5	–	310.2	y		d95		0.0002	s	−146						
Barium	Ba		137.3	s	3.5	725	1640	hyd	s	0	0	63	28	7.7	180	
	Ba								g	180	146	170	21			
	Ba^{2+}								g	1660						
	Ba^{2+}								aq	−538	−561	10				
acetate	$Ba(C_2H_3O_2)_2$		255.4	w	2.5	d150		78(3H$_2$O)	s	−1484						
bromide	$BaBr_2$	9	297.1	w	4.8	857	d		s	−757	−737	146	74	31	318	
	$\cdot 2H_2O$		333.2	w	3.6	d75		101	s	−1366	−1230	226				
carbonate	$BaCO_3$		197.3	w	4.4	d1360		0.002	s	−1216	−1138	112	85			
chloride	$BaCl_2$	6, 9	208.2	w	3.9	963	(1560)		s	−859	−810	124	75	17	333	
	$\cdot 2H_2O$		244.3	w	3.1	d113		37	s	−1460	−1296	203	162			
chromate	$BaCrO_4$		253.3	y	4.5	d		0.0003	s	−1446	−1345	159				
fluoride	BaF_2	6	175.3	w	4.9	1368	2137	0.16	s	−1207	−1157	96	71	23	389	
hydride	BaH_2	9	139.4	gy	4.2	d675		hyd	s	−179						
hydroxide	$Ba(OH)_2$		171.3	w	4.5	408			s	−945	−856	101	90	17	359	
	$\cdot 8H_2O$		315.5	w	2.2	d78		4.7	s	−3342	−2793	427				
iodide	BaI_2	9	391.1	w	5.1	711			s	−605	−601	165	77	27	276	
	$\cdot 2H_2O$		427.2	w	5.2	d99		220	s	−1217						
nitrate	$Ba(NO_3)_2$		261.3	w	3.2	592	d	10.1	s	−992	−797	214	151	25		

Substance	*Formula/Structure*		M / g mol^{-1}	*col.*	ρ / g cm^{-3}	t_m / °C	t_b / °C	*sol.* / g/100 g	*State*	$\Delta_f H^\circ$ / kJ mol^{-1}	$\Delta_f G^\circ$ / kJ mol^{-1}	S° / J K^{-1} mol^{-1}	C_p° / J K^{-1} mol^{-1}	$\Delta_{fus}H$ / kJ mol^{-1}	$\Delta_{sub}H$ / kJ mol^{-1}	p / 10^{-30} C m
Barium (cont.)																
nitride	Ba_3N_2	–	440.0	bn	4.8			hyd	s	−363	−292	152				
oxide	BaO	1	153.3	w	5.7	1918	(2000)	4	s	−554	−525	70	48	58	436	26.5
oxide, per-	BaO_2	–	169.3	w	5.0	450	d800	hyd	s	−634			67			
sulfate	$BaSO_4$		233.4	w	4.5	1580	d	0.00025	s	−1473	−1362	132	102	41		
sulfide	BaS	1	169.4	w	4.2	2230		hyd	s	−460	−456	78	49		510	36.2
Beryllium	Be		9.012	gy	1.9	1278	2471	i	s	0	0	9	16	12	324	
	Be								g	324	287	136	21			
	Be^{2+}								g	2993						
	Be^{2+}								aq	−383	−380	−130	69			
bromide	$BeBr_2$	–	168.8	w	3.5	*508	s490	s	s	−356	−337	100	66	10	126	
carbonate	$BeCO_3$		69.0						s	−1025						
	$\cdot 4H_2O$		141.1	w		d		i								
chloride	$BeCl_2$	8a	79.9	w	1.9	405	520	$72(4H_2O)$	s	−490	−445	83	65	8.7	132	
fluoride	BeF_2	9	47.0	w	2.0		s800	550	s	−1027	−979	53	52	4.7	233	
hydroxide	$Be(OH)_2$		43.0	w	1.9	d138		i	s	−903	−815	52	65		241	
iodide	BeI_2	–	262.8	w	4.3	480	590	hyd	s	−189	−187	120	69	21	125	
nitrate	$Be(NO_3)_2$		133.0													
	$\cdot 3H_2O$		187.1	w	1.6	60	d125	vs								
nitride	Be_3N_2	9	55.1	w	2.7	2200	d2240	hyd	s	−588	−533	34	64	129		
oxide	BeO	4	25.0	w	3.0	2507	(3900)	i	s	−609	−580	14	26	81	727	
sulfate	$BeSO_4$		105.1	w	2.4	d550			s	−1205	−1094	78	86			
	$\cdot 4H_2O$		177.1	w	1.7	d100		41	s	−2424	−2080	233	217			
sulfide	BeS	3	41.1	w	2.4	d		i	s	−264	−213	210	31			
Bismuth	Bi		209.0	s	9.8	271	1560	i	s	0	0	57	26	11	207	
	Bi								g	207	168	187	21			
	Bi^{3+}								g	5004						
	Bi^{3+}								aq		83					

Substance	*Formula/Structure*		M / g mol^{-1}	*col.*	ρ / g cm^{-3}	t_m / °C	t_b / °C	*sol.* / g/100 g	*State*	$\Delta_f H^\ominus$ / kJ mol^{-1}	$\Delta_f G^\ominus$ / kJ mol^{-1}	$S^\ominus$ / J K^{-1} mol^{-1}	$C_p^\ominus$ / J K^{-1} mol^{-1}	$\Delta_{fus}H$ / kJ mol^{-1}	$\Delta_{sub}H$ / kJ mol^{-1}	p / 10^{-30} C m
Bismuth (cont.)																
bromide	$BiBr_3$	0	448.7	y	5.7	218	453	hyd	s	−247			109	22		
chloride	$BiCl_3$	0	315.3	w	4.7	233	447	hyd	s	−379	−315	177	105	24	113	
hydroxide	$Bi(OH)_3$		260.0	w	5.0	d100		0.00014	s	−711						
iodide	BiI_3	8iv	589.7	bk	5.8	408	542	i	s	−151	−149	225	76	39		
nitrate	$Bi(NO_3)_3$		395.0													
	$\cdot 5H_2O$		485.1	w	2.8	d50		hyd								
oxide	Bi_2O_3	9	466.0	y	8.9	817	1890	i	s	−574	−494	151	114	28		
bismuthyl(III)	BiO^+		225.0						aq		−146					
oxychloride	$BiOCl$		260.4	w	7.7	d575		i	s	−367	−322	120				
sulfide	Bi_2S_3	8a	514.2	bn	7.4	d685		i	s	−143	−141	200	122	37		
Boron	B		10.81	y	2.3	2300	3660	i	s(β)	0	0	6	11	22	565	
	B								g	565	521	153	21			
acid, boric	H_3BO_3		61.8	w	1.4	d169		5.7	s	−1095	−970	90	81		100	
bromide	BBr_3	0	250.5	nil	2.6	−46	91	hyd	l	−240	−238	230	128		34(vap)	0
carbide	B_4C	9	55.3	bk	2.5	2350	>3500	i	s	−71	−71	27	53			
chloride	BCl_3	0	117.2	nil	1.4(l)	−107	12	hyd	g	−404	−389	290	63	2.1	26	0
fluoride	BF_3	0	67.8	nil		−127	−100	$0.31^{0°C}$	g	−1136	−1119	254	50	4.2		0
hydride (diborane(6))	B_2H_6	0	27.7	nil	0.4(l)	−166	−93	hyd	g	36	87	232	57	5		0
(tetraborane(10))	B_4H_{10}	0	53.3	nil	0.6(l)	−121	18	hyd	g	66						1.6
(pentaborane(9))	B_5H_9	0	63.1	nil	0.7	−47	60	hyd	l	43	172	184	151	6.1	37	7.1
(hexaborane(10))	B_6H_{10}	0	74.9	nil	0.7	−65	108	hyd	l	56					38(vap)	8.4
(decaborane(14))	$B_{10}H_{14}$	0	122.2	w	0.9	100	213	i	g	32	216	353	179	22	77	
iodide	BI_3	0	391.5	w	3.3(l)	50	210	hyd	g	71	21	349	71			
nitride	BN	8v, 3	24.8	w	2.2		s3000	i	s	−254	−228	15	20		902	
oxide	B_2O_3	9	69.6	w	2.5	450	2065	i	s	−1273	−1194	54	63	22	429	
sulfide	B_2S_3	9	117.8	w	1.6	310		hyd	s	−252	−248	92		48	308	

Substance	*Formula/Structure*		M / g mol^{-1}	*col.*	ρ / g cm^{-3}	t_m / °C	t_b / °C	*sol.* / g/100 g	*State*	$\Delta_f H^\circ$ / kJ mol^{-1}	$\Delta_f G^\circ$ / kJ mol^{-1}	S° / J K^{-1} mol^{-1}	C_p° / J K^{-1} mol^{-1}	$\Delta_{fus}H$ / kJ mol^{-1}	$\Delta_{sub}H$ / kJ mol^{-1}	p / 10^{-30} C m
Bromine	Br_2		159.8	r	3.1	−7	59	3.5	l	0	0	152	76	11	41	0
	Br_2								g	31	3	245	36			
	Br_2								aq	−3	4	130				
	Br		79.9						g	112	82	175	21			
	Br^-		79.9						g	−219	−239	163	21			
	Br^-		79.9						aq	−121	−104	83	−142			
chloride	$BrCl$	V	115.4	r		−66	d5	hyd	g	15	−1	240	35			1.7
fluoride, mono-	BrF	0	98.9	r		−33	d20		g	−94	−109	229	33			4.3
fluoride, tri-	BrF_3	V	136.9	y	2.8	9	126	hyd	l	−301	−240	−178	125		45(vap)	4.0
fluoride, penta-	BrF_5	0	174.9	nil	2.5	−61	41	hyd	l	−459	−352	225	100(g)	5.7	35	5.0
oxide, di-	BrO_2	–	111.9	y		d0			s	49						
bromate	BrO_3^-		127.9						aq	−67	19	162				
Cadmium	Cd		112.4	s	8.7	321	767	i	s	0	0	52	26	6.1	112	
	Cd								g	112	77	168	21			
	Cd^{2+}								g	2623						
	Cd^{2+}								aq	−76	−78	−73				
acetate	$Cd(C_2H_3O_2)_2$		230.5	w	2.3	256	d									
	$\cdot 2H_2O$		266.5	w	2.0	d130		156								
bromide	$CdBr_2$	8ii	272.2	y	5.2	567	863		s	−316	−296	137	77	33	151	
	$\cdot 4H_2O$		344.3	w		d36		112	s	−1493	−1248	316				
carbonate	$CdCO_3$		172.4	w	4.3	d310		i	s	−751	−669	92	82			
chloride	$CdCl_2$	8i	183.3	w	4.0	568	960		s	−391	−344	115	75	30	172	
	$\cdot H_2O$		201.3	w					s	−688	−587	168				
	$\cdot 2\frac{1}{2}H_2O$		228.4	w	3.3	d34		121	s	−1132	−944	227				
cyanide	$Cd(CN)_2$		164.4	w	2.2	d200		$1.7^{18°C}$	s	162	208	104				
fluoride	CdF_2	6	150.4	w	6.3	1100	1748	4.3	s	−700	−648	77	66	23		
hydroxide	$Cd(OH)_2$		146.4	w	4.8	d130		i	s	−561	−474	96				
iodide	CdI_2	8ii	366.2	g-y	5.7	387	796	86	s	−203	−201	161	80	21	138	
nitrate	$Cd(NO_3)_2$		236.4	w	3.6	350			s	−456				33		
	$\cdot 4H_2O$		308.5	w	2.4	59	132	159	s	−1649						

Substance	*Formula/Structure*		M g mol^{-1}	*col.*	ρ g cm^{-3}	t_m °C	t_b °C	*sol.* g/100 g	*State*	$\Delta_f H^\circ$ kJ mol^{-1}	$\Delta_f G^\circ$ kJ mol^{-1}	S° J K^{-1} mol^{-1}	C_p° J K^{-1} mol^{-1}	$\Delta_{fus}H$ kJ mol^{-1}	$\Delta_{sub}H$ kJ mol^{-1}	p 10^{-30} C m
Cadmium (cont.)																
nitride	Cd_3N_2	9	365.2	bk		d			s	162						
oxide	CdO	1	128.4	bn	8.1	1540		i	s	−258	−228	55	43		225	
sulfate	$CdSO_4$		208.5	w	4.7	1000			s	−933	−823	123	100			
	$\bullet 2\frac{2}{3}H_2O$		256.5	w	3.1	d41		77	s	−1729	−1465	230	213			
sulfide	CdS	3, 4	144.5	y	4.8	*1750	s980	i	s	−162	−156	65	55		210	
Caesium	Cs		132.9	s	1.9	28	669	hyd	s	0	0	85	32	2.1	76	
	Cs								g	76	49	176	21			
	Cs^+								g	458	427	170	21			
	Cs^+								aq	−258	−291	132	−10			
acetate	$CsC_2H_3O_2$		191.9	w		194		1020$^{21°C}$	aq	−744	−661	220				
bromide	CsBr	2	212.8	w	4.4	636	1300	123	s	−406	−391	113	53	24	197	
carbonate	Cs_2CO_3		325.8	w	4.2	d610		260$^{15°C}$	s	−1140	−1054	204	124			
chloride	CsCl	2	168.4	w	4.0	645	1303	190	s	−443	−415	101	52	16	203	34.8
fluoride	CsF	1	151.9	w	4.1	703	1251		s	−554	−526	93	51	22	194	26.3
	$\bullet 1\frac{1}{2}H_2O$		178.9			d180		370	s	−1014						
hydride	CsH	1	133.9	w	3.4	d		hyd	s	−54					170	
hydroxide	CsOH		149.9	y	3.7	315	990	327	s	−417	−371	99	68	11	157	23.7
iodide	CsI	2	259.8	w	4.5	626	1280	87	s	−347	−341	123	53	24	195	
nitrate	$CsNO_3$		194.9	w	3.7	414	d849	27	s	−506	−407	155		14		
oxide	Cs_2O	8i	281.8	o	4.3	d400		hyd	s	−346	−308	147	76		191	
oxide, super-	CsO_2	9	164.9	y	3.8	600	d	hyd	s	−286		146				
sulfate	Cs_2SO_4		361.9	w	4.2	1010		182	s	−1443	−1324	212	135	36		
sulfide	Cs_2S	9	297.9			520			s	−360						
	$\bullet 4H_2O$		369.9	w				vs								
Calcium	Ca		40.08	s	1.6	842	1484	hyd	s	0	0	42	26	8.5	178	
	Ca								g	178	144	155	21			
	Ca^{2+}								g	1926						
	Ca^{2+}								aq	−543	−553	−56				

Substance	*Formula/Structure*		M / g mol^{-1}	*col.*	ρ / g cm^{-3}	t_m / °C	t_b / °C	*sol.* / g/100 g	*State*	$\Delta_f H^\circ$ / kJ mol^{-1}	$\Delta_f G^\circ$ / kJ mol^{-1}	S° / J K^{-1} mol^{-1}	C_p° / J K^{-1} mol^{-1}	$\Delta_{fus}H$ / kJ mol^{-1}	$\Delta_{sub}H$ / kJ mol^{-1}	p / 10^{-30} C m
Calcium (cont.)																
acetate	$Ca(C_2H_3O_2)_2$		158.2	w	1.5	d > 160			s	−1479						
	$\cdot H_2O$		176.2	w		d		35	s	−1772						
	$\cdot 2H_2O$		194.2	w		d84			–							
bromide	$CaBr_2$	7*	199.9	w	3.4	742	1815		s	−683	−664	130	68	29	285	
	$\cdot 6H_2O$		308.0	w	2.3	d38		153	s	−2506	−2153	410				
carbide	CaC_2	9	64.1	gy	2.2	2300		hyd	s	−60	−65	70	63			
carbonate	$CaCO_3$		100.1	w	2.7	*1339	d899	0.0013	s(calcite)	−1207	−1129	93	82	53		
chloride	$CaCl_2$	7*	111.0	w	2.1	772	>1600		s	−796	−748	105	73	28	324	
	$\cdot 2H_2O$		147.0	w	1.8				s	−1403						
	$\cdot 6H_2O$		219.1	w	1.7	27	d30	83	s	−2608						
chromate	$CaCrO_4$		156.1													
	$\cdot 2H_2O$		192.1	y	2.5	d200		17								
citrate	$Ca_3(C_6H_5O_7)_2$		498.4													
	$\cdot 4H_2O$		570.5	w		d120		0.1								
fluoride	CaF_2	6	78.1	w	3.2	1418	2533	0.0016	s	−1220	−1167	69	67	30	438	
hydride	CaH_2	9	42.1	w	1.9	*816	d600	hyd	s	−186	−147	42	37			
hydroxide	$Ca(OH)_2$		74.1	w	2.3	d580		0.12	s	−986	−898	83	87		442	
hypochlorite	$Ca(OCl)_2$		143.0	w	2.4	d100		hyd								
iodide	CaI_2	8ii	293.9	w	4.0	784	1100	210	s	−533	−529	142		42	261	
nitrate	$Ca(NO_3)_2$		164.1	w	2.5	561			s	−938	−743	193	149	21		
	$\cdot 4H_2O$		236.2	w	1.9	43	d132	138	s	−2132	−1713	375				
nitride	Ca_3N_2	9	148.2	bn	2.6	1195		hyd	s	−431			114			
oxalate	CaC_2O_4		128.1	w	2.2	d			s	−1361						
	$\cdot H_2O$		146.1	w	2.2	d200		0.0007	s	−1675	−1514	157	153			
oxide	CaO	1	56.1	w	3.3	2927	3500	hyd	s	−635	−603	38	42	79		
phosphate	$Ca_3(PO_4)_2$		310.2	w	3.1	1670		0.002	s	−4121	−3885	236	228			
hydrogen	$CaHPO_4$		136.1						s	−1814	−1681	111	110			
	$\cdot 2H_2O$		172.1	w	2.3	d109		0.014	s	−2404	−2155	189	197			
dihydrogen	$Ca(H_2PO_4)_2$		234.1			d203			s	−3105						
	$\cdot H_2O$		252.1	w	2.2	d109		$1.8^{30\,°C}$	s	−3410	−3058	260	259			

Substance	*Formula/Structure*		M g mol⁻¹	*col.*	ρ g cm⁻³	t_m °C	t_b °C	*sol.* g/100 g	*State*	$\Delta_f H^\circ$ kJ mol⁻¹	$\Delta_f G^\circ$ kJ mol⁻¹	S° J K⁻¹ mol⁻¹	C_p° J K⁻¹ mol⁻¹	$\Delta_{fus}H$ kJ mol⁻¹	$\Delta_{sub}H$ kJ mol⁻¹	p 10⁻³⁰ C m
Calcium (cont.)																
phosphide	Ca_3P_2		182.2	r-bn	2.5	(1600)		hyd	s	−506	−481	124				
sulfate	$CaSO_4$		136.1	w	3.0	1400			s	−1434	−1322	107	100	28		
	$\cdot\frac{1}{2}H_2O$		145.1	w		d163			s	−1577	−1437	131	119			
	$\cdot 2H_2O$		172.2	w	2.3	d128		0.21	s	−2023	−1797	194	186			
sulfide	CaS	1	72.1	W	2.6	2525		hyd	s	−482	−477	56	47			
Carbon	C (graphite)		12.01	bk	2.3	*3974	s3930	i	s(graphite)	0	0	6	9	105	716	
	C (diamond)		12.01	nil	3.5	>3550		i	s(diamond)	2	3	2	6		715	
	C								g	717	671	158	21			
acid, hydrocyanic	HCN		27.0	nil	0.7	−13	26	vs	l	109	125	113	71	8	35	9.9
	HCN								g	135	125	202	36			
	HCN								aq	151	172	94				
bromide, tetra-	CBr_4	V	331.6	w	3.4	92	190	0.02	s	19	48	213	144	4.0	60	0
chloride, tetra-	CCl_4	V	153.8	nil	1.6	−23	77	0.08	l	−128	−58	216	132	2.5	35	0
oxy-, (phosgene)	$COCl_2$		98.9	nil	1.4	−128	8	hyd	g	−219	−205	284	58	5.7		3.9
fluoride, tetra-	CF_4	V	88.0	nil	1.9	−184	−128	0.0018	g	−933	−888	262	61	0.7		0
iodide, tetra-	CI_4	V	519.6	r	4.2$^{20°C}$	171	d	i	g			392	96			0
nitride (cyanogen)	C_2N_2	V	52.0	nil	1.0(l)	−28	−21	1.2	g	309	297	242	57	8	28	0
cyanogen bromide	BrCN		105.9	w	2.0	52	61	s	g	186	165	248	47		46	9.8
cyanogen chloride	ClCN		61.5	nil	1.2	−6	13	s	g	138	131	236	45	11	37	9.4
cyanogen fluoride	FCN		45.0	nil		−82	−46		g			225	42			7.2
cyanogen iodide	ICN		152.9	w	1.8	*146	s45	3.85	g	226	197	257	48		59	12.4
oxide, mon-	CO	V	28.0	nil	0.8(l)	−205	−192	0.0026	g	−111	−137	198	29	0.83		0.4
oxide, di-	CO_2	0	44.0	nil	1.1(l), 1.6(s)	−57*	s−78	0.145	g	−394	−394	214	37	9.0	25	0.0
	CO_2								aq	−413	−386	119				
carbonate	CO_3^{2-}		60.0						aq	−675	−528	−50				
hydrogen carbonate	HCO_3^-		61.0						aq	−690	−587	98				
acetate	$CH_3CO_2^-$		59.0						aq	−486	−369	87	−6			
sulfide, di-	CS_2	V	76.1	nil	1.3	−111	46	0.17	l	90	65	151	76	4.4	32	0.0

Substance	*Formula/Structure*		M / g mol^{-1}	*col.*	ρ / g cm^{-3}	t_m / °C	t_b / °C	*sol.* / g/100 g	*State*	$\Delta_f H^\circ$ / kJ mol^{-1}	$\Delta_f G^\circ$ / kJ mol^{-1}	S° / J K^{-1} mol^{-1}	C_p° / J K^{-1} mol^{-1}	$\Delta_{fus}H$ / kJ mol^{-1}	$\Delta_{sub}H$ / kJ mol^{-1}	p / 10^{-30} C m
Carbon (cont.)																
thiocyanate	SCN^-		58.1						aq	76	93	144	−40			
Cerium	Ce		140.1	gy	6.8	795	3433	i	s	0	0	72	26	5	423	
	Ce								g	423	385	192	23			
	Ce^{3+}								g	3971						
	Ce^{3+}								aq	−696	−672	−205				
	Ce^{4+}								g	7523						
	Ce^{4+}								aq	−537	−504	−301				
(IV) ammonium nitrate	$(NH_4)_2Ce(NO_3)_6$		548.2	o				141								
(III) chloride	$CeCl_3$	9	246.5	w	4.0	848	1727	100	s	−1054	−978	151	87	54	326	
	$\cdot 7H_2O$		372.6	y	3.9	d90		vs	s	−3169						
(III) nitrate	$Ce(NO_3)_3$		326.1						s	−1226						
	$\cdot 6H_2O$		434.2	r		d150		176	s	−3051						
(III) oxide	Ce_2O_3	9	328.2	gy	6.9	2230	3730	i	s	−1796	−1706	151	115			
(IV) oxide	CeO_2	6	172.1	w	7.1	(2600)		i	s	−1089	−1025	62	62			
(IV) sulfate	$Ce(SO_4)_2$		332.2	y	3.9	d195		hyd	s	−2343						
	$\cdot 4H_2O$		404.3	y	3.9	d180		hyd								
Chlorine	Cl_2		70.90	y	1.6(l)	−101	−34	0.64	g	0	0	223	34	6.4	24	0
	Cl_2								aq	−23	7	121				
	Cl		35.45						g	121	105	165	22			
	Cl^-		35.45						g	−234	−240	153	21			
	Cl^-		35.45						aq	−167	−131	57	−136			
acid, chloric	$HClO_3$		84.5						aq	−104	−8	162				
	$\cdot 7H_2O$		210.6	nil	1.3	<−20	d40	vs	l	−41						
perchloric	$HClO_4$		100.5	nil	1.8(l)	−112	d	vs	s	−382						
	$\cdot H_2O$		118.5	w	1.9	50	d100	vs	l	−678						
	$\cdot 2H_2O$		136.5	nil	1.7	−18	200	vs	aq	−128	−8	184				
fluoride, mono-	ClF	V	54.5	nil	1.6(l)	−156	−101	hyd	g	−54	−56	218	32			2.9
fluoride, tri-	ClF_3	0	92.5	nil	1.8(l)	−76	12	hyd	g	−163	−123	282	64		26(vap)	2.0

Substance	*Formula/Structure*		M / g mol^{-1}	*col.*	ρ / g cm^{-3}	t_m / °C	t_b / °C	*sol.* / g/100 g	*State*	$\Delta_f H^\circ$ / kJ mol^{-1}	$\Delta_f G^\circ$ / kJ mol^{-1}	S° / J K^{-1} mol^{-1}	C_p° / J K^{-1} mol^{-1}	$\Delta_{fus}H$ / kJ mol^{-1}	$\Delta_{sub}H$ / kJ mol^{-1}	p / 10^{-30} C m
Chlorine (cont.)																
hypochlorite	ClO^-		51.5						aq	−107	−37	42				
(di-)oxide, mon-	Cl_2O	V	86.9	y	3.0(l)	−20	d4	hyd	g	80	98	266	45			
oxide, di-	ClO_2	V	67.5	y	1.6(l)	−60	d10	11	g	102	120	257	42			6.0
(di-)oxide, hepta-	Cl_2O_7	V	182.9	nil	1.8(l)	−92	82	hyd	g	272						
Chromium	Cr		52.00	gy	7.2	1857	2672	i	s	0	0	24	23	21	397	
	Cr								g	397	352	175	21			
	Cr^{2+}								g	2656						
	Cr^{3+}								aq	−144						
	Cr^{3+}								g	5648						
(II) chloride	$CrCl_2$	7*	122.9	w	2.9	814	1300	vs	s	−395	−356	115	71	32	267	
(III) chloride	$CrCl_3$	8iii	158.4	r	2.8	(1150)	d		s	−556	−486	123	92		238	
	$\bullet 6H_2O$		266.5	v	1.8	83		246								
chloride, chromyl	CrO_2Cl_2		154.9	r	1.9	−96	117	hyd	l	−580	−511	222	(g)85		41(vap)	
(III) nitrate	$Cr(NO_3)_3$		238.0													
	$\bullet 9H_2O$		400.2	v	1.8	60	d100	79	s				457			
(III) oxide	Cr_2O_3	9	152.0	g	5.2	2266	3000	i	s	−1140	−1058	81	119	130		
oxide, di-	CrO_2	7	84.0	bk	4.9	d300		i	s	−598	−545	51				
(VI) oxide	CrO_3	8a	100.0	r	2.7	196	d250	169	s	−590	−513	72	69	14	204	
chromate	CrO_4^{2-}		116.0						aq	−881	−728	50				
dichromate	$Cr_2O_7^{2-}$		216.0						aq	−1490	−1301	262				
(III) potassium sulfate	$CrK(SO_4)_2$		283.2						s	−2351						
	$\bullet 12H_2O$		499.4	r	1.8	89	d100	12.5	s	−5777						
(III) sulfate	$Cr_2(SO_4)_3$		392.2	v	3.0			64(16H_2O)	s	−2911	−2578	259	282			
Cobalt	Co		58.93	gy	8.9	1495	2870	i	s	0	0	30	25	16	425	
	Co								g	425	380	180	23			
	Co^{2+}								g	2844						
	Co^{2+}								aq	−58	−54	−113				
	Co^{3+}								g	6083						
	Co^{3+}								aq	92	134	−305				

Substance	*Formula/Structure*		*M* g mol⁻¹	*col.*	*ρ* g cm⁻³	t_m °C	t_b °C	*sol.* g/100 g	*State*	$\Delta_f H^\circ$ kJ mol⁻¹	$\Delta_f G^\circ$ kJ mol⁻¹	S° J K⁻¹ mol⁻¹	C_p° J K⁻¹ mol⁻¹	$\Delta_{fus}H$ kJ mol⁻¹	$\Delta_{sub}H$ kJ mol⁻¹	*p* 10⁻³⁰ C m
Cobalt (cont.)																
acetate	$Co(C_2H_3O_2)_2$		177.0	p												
	$\cdot 4H_2O$		249.1	r	1.7	d140		s								
bromide	$CoBr_2$	8ii	218.7	g	4.9	678			s	−221	−207	134				
	$\cdot 6H_2O$		326.8	r	2.5	d47		119	s	−2020						
carbonate	$CoCO_3$		118.9	r	4.1	d		i	s	−713	−637	89				
hydroxide	$2CoCO_3 \cdot 3Co(OH)_2 \cdot H_2O$[(b)]		534.7					i								
chloride	$CoCl_2$	8i	129.8	b	3.4	740	1049		s	−313	−270	109	78	59	219	
	$\cdot 6H_2O$		237.9	r	1.9	d87		56	s	−2115	−1725	343				
hydroxide	$Co(OH)_2$		93.0	p	3.6	d		0.0003	s(pink)	−540	−454	79				
									s(blue)		−450					
iodide	CoI_2	8ii	312.7	bk	5.6	515		203($1H_2O$)	s	−89	−91	153				
nitrate	$Co(NO_3)_2$		182.9	r	2.5	d100			s	−420						
	$\cdot 6H_2O$		291.0	r	1.9	d55		102	s	−2211			452	37		
(II) oxide	CoO	1	74.9	bk	6.5	1795		i	s	−238	−214	53	55			
(III) oxide	Co_2O_3		165.9	bk	5.2	d895		i								
sulfate	$CoSO_4$		155.0	b	3.7	d735			s	−888	−782	118	138			
	$\cdot 6H_2O$		263.1	r	2.0	d95			s	−2684	−2236	368	353			
	$\cdot 7H_2O$		281.1	r	1.9	97	d420	38	s	−2980	−2474	406	390			
sulfide	CoS	5	91.0	r	5.5	>1100		i	s	−83			47			
thiocyanate	$Co(SCN)_2$		175.1					8	s	101						
	$\cdot 3H_2O$		229.1	v		d105		s								
Copper	Cu		63.55	r	9.0	1085	2572	i	s	0	0	33	24	13	338	
	Cu								g	337	298	166	21			
	Cu^+								g	1090	1027	222	21			
	Cu^+								aq	72	50	41				
	Cu^{2+}								g	3054		179				
	Cu^{2+}								aq	65	65	−98				

[(b)] ($2CoCO_3 \cdot 3Co(OH)_2 \cdot H_2O$) – approximate formula of the precipitated carbonate.

Substance	*Formula/Structure*		M g mol^{-1}	*col.*	ρ g cm^{-3}	t_m °C	t_b °C	*sol.* g/100 g	*State*	$\Delta_f H^\circ$ kJ mol^{-1}	$\Delta_f G^\circ$ kJ mol^{-1}	S° J K^{-1} mol^{-1}	C_p° J K^{-1} mol^{-1}	$\Delta_{fus}H$ kJ mol^{-1}	$\Delta_{sub}H$ kJ mol^{-1}	p 10^{-30} C m
Copper (cont.)																
(II) acetate	$Cu(C_2H_3O_2)_2$		181.6	g	1.9				s	−893						
	$\cdot H_2O$		199.7	g	1.9	115	d240	7.3	s	−1189						
(I) bromide	CuBr	3	143.5	w	5.0	504	1345	i	s	−105	−101	96	55	10		
(II) bromide	$CuBr_2$	8a	223.3	bk	4.8	498	900	126	s	−142	−127	134				
(II) carbonate hydroxide	$CuCO_3\cdot Cu(OH)_2$[c]		221.1	g	4.0	d200		i	s	−1051	−894	186				
(I) chloride	CuCl	3	99.0	w	4.1	430	1490	0.02	s	−137	−120	86	49	7	242	
(II) chloride	$CuCl_2$	8a	134.4	y	3.4	d>300			s	−220	−176	108	72	20		
	$\cdot 2H_2O$		170.5	g	2.5	d100		77	s	−821	−656	167				
(I) cyanide	CuCN		89.6	w	2.9	473	d	0.00026	s	96	111	85				
(II) fluoride	CuF_2	7*	101.5	w	4.2	d950		0.07	s	−539	−492	77	66	55	262	
	$\cdot 2H_2O$		137.6	b	2.9	d130			s		−981					
(II) hydroxide	$Cu(OH)_2$		97.6	b	3.4	d160		i	s	−450	−373	108	95			
(I) iodide	CuI	3	190.4	w	5.6	605	1290	0.008$^{18°C}$	s	−68	−69	97	54	11		
(II) nitrate	$Cu(NO_3)_2$		187.6						s	−303						
	$\cdot 3H_2O$		241.6	b	2.3	115	d170		s	−1217						
	$\cdot 6H_2O$		295.6	b	2.1	d26		150	s	−2111						
(I) oxide	Cu_2O	9	143.1	r	6.0	1235	d1800	i	s	−169	−146	93	64	57		
(II) oxide	CuO	9	79.5	bk	6.4	1326		i	s	−157	−130	43	42	12	402	
(II) sulfate	$CuSO_4$		159.6	w	3.6	d560			s	−771	−662	109	100			
	$\cdot 5H_2O$		249.7	b	2.3	d110		22	s	−2280	−1880	300	280			
(I) sulfide	Cu_2S	9	159.1	bk	5.6	1130		i	s	−80	−86	121	76	11		
(II) sulfide	CuS	9	95.6	bk	4.6	d220		i	s	−53	−54	67	48		403	
(I) thiocyanate	CuSCN		121.6	w	2.8	1084		0.0005$^{18°C}$	s		70					
Fluorine	F_2		38.00	y	1.51(l)	−220	−188	hyd	g	0	0	203	31	0.51		0
	F		19.00						g	79	62	159	23			
	F^-		19.00						g	−255	−262	146	21			
	F^-		19.00						aq	−335	−281	−14	−107			
	HF_2^-		39.0						aq	−650	−578	92				

[c] ($CuCO_3\cdot Cu(OH)_2$) – approximate formula of the precipitated carbonate.

Substance	*Formula*	*Structure*	M / g mol^{-1}	*col.*	ρ / g cm^{-3}	t_m / °C	t_b / °C	*sol.* / g/100 g	*State*	$\Delta_f H^\circ$ / kJ mol^{-1}	$\Delta_f G^\circ$ / kJ mol^{-1}	S° / J K^{-1} mol^{-1}	C_p° / J K^{-1} mol^{-1}	$\Delta_{fus}H$ / kJ mol^{-1}	$\Delta_{sub}H$ / kJ mol^{-1}	p / 10^{-30} C m
Fluorine (cont.)																
oxide	F_2O See OF_2	V														
Gallium	Ga		69.72	S	5.9	30	2403	i	s	0	0	41	26	5.6	277	
	Ga								l	6			28			
	Ga								g	277	239	169	25			
	Ga^{3+}								g	5817						
	Ga^{3+}								aq	−212	−159	−331				
(III) bromide	$GaBr_3$	0(d)	309.4	w	3.7	122	279	s	s	−387	−360	180		12.1	94	
(II) chloride	$GaCl_2$	9	140.6	w	2.7	164	535	hyd								
(III) chloride	$GaCl_3$	0(d)	176.1	w	2.5	78	201	vs	s	−525	−455	142		11	77	
(III) fluoride	GaF_3	9	126.7	w	4.5		s950	0.002	s	−1163	−1085	84				
hydride	GaH	–	70.7						g	220	194	195	29			
hydride (digallane)	Ga_2H_6	–	145.5	nil		−21	d>130	hyd								
hydroxide	$Ga(OH)_3$		120.7	w		d440		i	s	−964	−832	100				
(III) iodide	GaI_3	0(d)	450.4	y	4.2	212	340	hyd	s	−239	−236	204		16	97	0
(III) nitride	GaN	4	83.7	gy	6.1		s800	i	s	−110	−78	30	41		286	
(I) oxide	Ga_2O	–	155.4	bn	4.8	*>660	s>500	i	s	−356					268	
(III) oxide	Ga_2O_3	9	187.4	w	5.9	1795		i	s	−1089	−998	85	92			
(III) sulfide	Ga_2S_3	9	235.6	y	3.6	1250		hyd	s	−514						
Germanium	Ge		72.63	s	5.3	937	2830	i	s	0	0	31	23	32	372	
	Ge								g	372	331	168	31			
(II) chloride	$GeCl_2$	–	143.5	w		d450		hyd								
(IV) chloride	$GeCl_4$	–	214.4	nil	1.9	−50	84	hyd	l	−532	−463	246	96(g)		36(vap)	0
hydride (germane)	GeH_4	0	76.7	nil	1.5(l)	−165	−88	i	g	91	113	217	45	1		0
hydride (digermane)	Ge_2H_6	0	151.3	nil	2	−109	29	hyd	l	137					25(vap)	
(IV) nitride	Ge_3N_4	9	273.9	w	5.2	d450		i	s	−63	33	155				
(II) oxide	GeO	–	88.6	bk			s710	0.002	s	−262	−237	50			216	11.0
(IV) oxide	GeO_2	7	104.6	w	4.2	1115	1200	0.45	s	−580	−521	40	50	44		

Substance	*Formula/Structure*		M / g mol^{-1}	*col.*	ρ / g cm^{-3}	t_m / °C	t_b / °C	*sol.* / g/100 g	*State*	$\Delta_f H^\circ$ / kJ mol^{-1}	$\Delta_f G^\circ$ / kJ mol^{-1}	S° / J K^{-1} mol^{-1}	C_p° / J K^{-1} mol^{-1}	$\Delta_{fus}H$ / kJ mol^{-1}	$\Delta_{sub}H$ / kJ mol^{-1}	p / 10^{-30} C m
Germanium (cont.)																
(II) sulfide	GeS	1*	104.7	o	4.0	*530	s430	0.25	s	−69	−71	71		21	161	6.7
(IV) sulfide	GeS_2	9	136.8	w	2.9		s>600	0.45	s	−157	−155	87				
Gold	Au		197.0	y	19.3	1064	2856	i	s	0	0	47	25	13		
	Au								g	366	326	180	21			
acid, chloroauric (III)	$HAuCl_4$		339.8	y				vs								
(I) chloride	AuCl	8a	232.4	y	7.6	d289		i	s	−35	−15	93				
(III) chloride	$AuCl_3$	0(d)	303.3	r	4.7	d160		$68^{20°C}$	s	−118	−48	148				
(III) hydroxide	$Au(OH)_3$		248.0	bn		d100		i	s	−425	−317	189				
(III) oxide	Au_2O_3	9	441.9	bn		d150		i	s	−3	80					
Hydrogen	H_2		2.016	nil	0.1(l)	−259	−253	0.00015	g	0	0	131	29	0.12		0
(deuterium)	$^2H_2$ or D_2		4.028	nil	0.2(l)	−255	−250	i	g	0	0	145	29			0
	H								g	218	203	115	21			
	H^+								g	1537	1517	109	21			
	H^+								aq	0	0	0	0			
	H^-								g	139	132	109	21			
bromide	HBr	V	80.9	nil	2.8(l)	−87	−67	186	g	−36	−53	199	29	2.4	15	2.7
const. boil. 47%	HBr			nil	1.5	−11	126		aq	−122	−104	82	−142			
chloride	HCl	V	36.5	nil	1.2(l)	−114	−85	70	g	−92	−95	187	29	2	11	3.7
const. boil. 20%	HCl			nil	1.1		110		aq	−167	−131	56	−136			
fluoride	HF	V	20.0	nil	1.0(l)	−83	20	vs	g	−273	−275	174	29	4.6	31	6.1
const. boil. 35%	HF			nil			120		aq	−333	−279	−14	−107			
iodide	HI	V	127.9	nil	2.8(l)	−51	−35	vs	g	26	2	207	29	2.9	20	1.5
const. boil. 57%	HI			nil	1.7		127		aq	−55	−52	111	−142			
oxide (water)	H_2O	0	18.0	nil	1.0	0	100		s	$-293^{0°C}$	$-240^{0°C}$	$41^{0°C}$	$39^{0°C}$			
	H_2O								l	−286	−237	70	75	6.0	50	6.2
	H_2O								g	−242	−229	189	34			
	OH^-		17.0						aq	−230	−157	−11	−149			
(heavy water)	^2H_2O or D_2O	0	20.0	nil	1.1	4	101	∞	l	−295	−243	76	84	6	52	6.2

Substance	*Formula/Structure*		M	*col.*	ρ	t_m	t_b	*sol.*	*State*	$\Delta_f H^\circ$	$\Delta_f G^\circ$	S°	C_p°	$\Delta_{fus} H$	$\Delta_{sub} H$	p
			g mol^{-1}		g cm^{-3}	°C	°C	g/100 g		kJ mol^{-1}	kJ mol^{-1}	J K^{-1} mol^{-1}	J K^{-1} mol^{-1}	kJ mol^{-1}	kJ mol^{-1}	10^{-30} C m
Hydrogen (cont.)																
oxide, per-	H_2O_2	0	34.0	nil	1.4	−0.4	150	∞	l	−188	−120	110	89	12	64	7.3
oxide, per- (urea)	$H_2O_2•CO(NH_2)_2$[d]		94.1	w		d60		s								
selenide	H_2Se	V	81.0	nil	2.0	−64	−41	0.68	g	30	16	219	35	2.5		0.8
sulfide	H_2S	V	34.1	nil		−86	−60	0.33	g	−21	−34	206	34	2.4	16	3.2
	S^{2-}		32.1						aq	33	86	−15				
	SH^-		33.1						aq	−16	12	67				
telluride	H_2Te	–	129.6	nil	2.6(1)	−49	−2	s	g	100	85	229	36	6.70		<0.7
Indium	In		114.8	s	7.3	157	2080	i	s	0	0	58	27	3.3	243	
	In								g	243	209	174	21			
	In^{3+}								g	5346						
	In^{3+}								aq	−105	−98	−151				
(I) chloride	InCl	1*, 8v	150.3	r	4.2	225	608	hyd	s	−186	−164	95		13	111	12.6
(III) chloride	$InCl_3$	8iii	221.2	w	4.0	*586	s500	vs	s	−537	−462	141			163	
(I) hydride	InH	–	115.8						g	215	190	208	30			
(III) nitride	InN	4	128.8	bn	6.9	1100			s	−138	−105	44				
(III) oxide	In_2O_3	9	277.6	y	7.2	1912		i	s	−926	−831	104	92			
(III) sulfide	In_2S_3	9	325.8	y	4.9	1050		i	s	−427	−413	164	118			
Iodine	I_2		253.8	v	4.9	114	184	0.034	s	0	0	116	54	16	62	0
	I_2								g	62	19	261	37			
	I_2								aq	23	16	137				
	I		126.9						g	107	70	181	21			
	I^-		126.9						g	−195	−221	169	21			
	I^-		126.9						aq	−57	−52	106	−142			
	I_3^-		380.7						aq	−51	−51	239				
acid, iodic	HIO_3		175.9	w	4.6	d110		263	s	−230						
	IO_3^-		174.9						aq	−221	−128	118				
bromide, mono-	IBr	0	206.8	gy	4.4	40	d116	hyd	g	41	4	259	36		51	2.4
chloride, mono-	ICl	8a	162.4	r	3.2	27	d97	hyd	g	18	−5	248	36	11	53	2.0
chloride, tri-	ICl_3	0(d)	233.3	o	3.2	*101	d77	hyd	s	−89	−22	167				

[d] ($H_2O_2•CO(NH_2)_2$): H_2O_2 is stabilised as solid hydrogen peroxide urea $H_2O_2•CO(NH_2)_2$ (36%H_2O_2), known also as carbamide peroxide, percarbamide, perhydrol and perhydrit tablets.

Substance	*Formula/Structure*		M g mol^{-1}	*col.*	ρ g cm^{-3}	t_m °C	t_b °C	*sol.* g/100 g	*State*	$\Delta_f H^\circ$ kJ mol^{-1}	$\Delta_f G^\circ$ kJ mol^{-1}	S° J K^{-1} mol^{-1}	C_p° J K^{-1} mol^{-1}	$\Delta_{fus}H$ kJ mol^{-1}	$\Delta_{sub}H$ kJ mol^{-1}	p 10^{-30} C m
Iodine (cont.)																
fluoride, penta-	IF_5	V	221.9	nil	3.2	9	100	hyd	g	−822	−752	328	99	16	58	7.3
fluoride, hepta-	IF_7	0	259.9	nil	2.8	*6	s5	hyd	g	−944	−818	346	136		31	
(di-)oxide, tetra-	I_2O_4	–	317.8	y	4.2	d75		hyd								
(di-)oxide, penta-	I_2O_5	V	333.8	w	4.8	d300		hyd	s	−158						
Iron	Fe		55.85	s	7.9	1535	2750	i	s	0	0	27	25	14	416	
	Fe								g	416	371	180	26			
	Fe^{2+}								g	2750						
	Fe^{2+}								aq	−89	−79	−138				
	Fe^{3+}								g	5713						
	Fe^{3+}								aq	−49	−5	−316				
(II) ammonium sulfate	See Ammonium, page 37															
(III) ammonium sulfate	See Ammonium, page 37															
arsenide	FeAs	5	130.8	w	7.8	1020		i								
(II) bromide	$FeBr_2$	8ii	215.7	g	4.6	d684		119	s	−250	−238	141		50	204	
	$\cdot 6H_2O$		323.8	g	4.6	d27										
(III) bromide	$FeBr_3$	8iv	295.6	r	4.5	d		s	s	−268	−243	174		50	144	
	$\cdot 6H_2O$		403.7	g		27										
carbide	Fe_3C	9	179.6	gy	7.7	1837		i	s	25	20	105	106	51		
(II) carbonate	$FeCO_3$		115.9	gy	3.8	d		$0.072^{18\,°C}$	s	−741	−667	93	82			
(II) chloride	$FeCl_2$	8i	126.8	y	3.2	674	1024		s	−342	−302	118	77	43	193	
	$\cdot 4H_2O$		198.8	b-g	1.9			65	s	−1549						
(III) chloride	$FeCl_3$	8iv	162.2	bn	2.9	306	d315		s	−399	−334	142	97	43	145	
	$\cdot 6H_2O$		270.3	bn		d37		50	s	−2224						
(II) hydroxide	$Fe(OH)_2$		89.9	g	3.4	d		i	s	−569	−487	88	97		197	
(III) hydroxide	$Fe(OH)_3$		106.9	bn	3.9	d500		i	s	−823	−697	107	102			
(II) iodide	FeI_2	8ii	309.7	gy	5.3	177		s	s	−116	−124	170		45	174	
	$\cdot 4H_2O$		381.7	gy	2.9	d90		vs								

Substance	*Formula/Structure*		M g mol^{-1}	*col.*	ρ g cm^{-3}	t_m °C	t_b °C	*sol.* g/100 g	*State*	$\Delta_f H^\circ$ kJ mol^{-1}	$\Delta_f G^\circ$ kJ mol^{-1}	S° J K^{-1} mol^{-1}	C_p° J K^{-1} mol^{-1}	$\Delta_{fus}H$ kJ mol^{-1}	$\Delta_{sub}H$ kJ mol^{-1}	p 10^{-30} C m
Iron (cont.)																
(III) nitrate	$Fe(NO_3)_3$		241.9													
	$\cdot 9H_2O$		404.0	v	1.7	47	d100	87	s	−3285						
(II) oxalate	FeC_2O_4		143.9													
	$\cdot 2H_2O$		179.9	y	2.3	d150		0.008	s	−1482						
(II) oxide	FeO	1	71.8	bk	5.7	1360	d3414	i	s	−272	−251	61	50	24	523	
(II/III) oxide	Fe_3O_4	9	231.5	bk	5.2	1597		i	s	−1118	−1015	146	143	138		
(III) oxide	Fe_2O_3	9	159.7	r	5.2	1565		i	s	−824	−742	87	104			
(III) phosphate	$FePO_4$		150.8						s	−1297						
	$\cdot 2H_2O$		186.8	p	2.7	d		i	s	−1888	−1658	171	181			
(II) sulfate	$FeSO_4$		151.9	w	3.7	d			s	−928	−821	108	101			
	$\cdot 7H_2O$		278.0	g	1.9	d64		29.5	s	−3015	−2510	409	394			
(III) sulfate	$Fe_2(SO_4)_3$		399.9	y	3.1	d480		hyd	s	−2582	−2262	308				
	$\cdot xH_2O$			y												
(II) sulfide	FeS	5	87.9	bk	4.8	1195	d	0.0006	s	−100	−100	60	51	32		
sulfide (pyrite)	FeS_2	9	120.0	y	5.0	1171	d	0.0005	s(pyrite)	−172	−160	53	62			
sulfide (marcasite)	FeS_2	9	120.0	y	4.9	d450	d	0.0005	s(marcasite)	−167	−156	54	62			
Krypton	Kr		83.80	nil	2.2	−157	−152	0.020	g	0	0	164	21	1.6		0
fluoride, di-	KrF_2	V	121.8	w	3.2		s−60	hyd	g	60					41	0
Lanthanum	La		138.9	s	6.2	920	3470	hyd	s	0	0	57	27	6.2		
	La								g	431	394	182	23			
	La^{3+}								g	3905						
	La^{3+}								aq	−707	−684	−218	−13			
chloride	$LaCl_3$		245.3	w	3.8	852	1812	97	s	−1071						
	$\cdot 7H_2O$		371.4	w		d91			s	−3179	−2713	463	431		330	
nitrate	$La(NO_3)_3$		324.9						s	−1254						
	$\cdot 6H_2O$		433.0	w		40	d126	147	s	−3064						
oxide	La_2O_3		325.8	w	6.5	2320	4200	0.0001	s	−1794	−1706	127	109			

Substance	*Formula/Structure*		M g mol^{-1}	*col.*	ρ g cm^{-3}	t_m °C	t_b °C	*sol.* g/100 g	*State*	$\Delta_f H^\circ$ kJ mol^{-1}	$\Delta_f G^\circ$ kJ mol^{-1}	S° J K^{-1} mol^{-1}	C_p° J K^{-1} mol^{-1}	$\Delta_{fus}H$ kJ mol^{-1}	$\Delta_{sub}H$ kJ mol^{-1}	p 10^{-30} C m
Lead	Pb		207.2	gy	11.3	327	1740	i	s	0	0	65	26	4.8	195	
	Pb								g	195	162	175	21			
	Pb^{2+}								g	2373						
	Pb^{2+}								aq	1	−24	18				
(II) acetate	$Pb(C_2H_3O_2)_2$		325.3	w	3.2	280	d		s	−964						
	•$3H_2O$		379.3	w	2.6	d75		55	s	−1852						
(IV) acetate	$Pb(C_2H_3O_2)_4$		443.4	w	2.2	175		hyd								
arsenate	$Pb_3(AsO_4)_2$		899.4	w	7.8	1042	d	i								
hydrogen	$PbHAsO_4$		347.1	w	5.9	d280		i								
(II) azide	$Pb(N_3)_2$		291.2	nil	4.7	d350		$0.025^{20°C}$	s	478	625	148				
bromide	$PbBr_2$	9	367.0	w	6.7	373	914	0.97	s	−279	−262	161	80	19	173	
carbonate	$PbCO_3$		267.2	w	6.6	d340		0.00017	s	−699	−626	131	87			
hydroxide	$(PbCO_3)_2$•$Pb(OH)_2$ [e]		775.6	w	6.1	d400		0.00016								
(II) chloride	$PbCl_2$	9	278.1	w	5.9	501	950	1.08	s	−359	−314	136	77	24	185	
(IV) chloride	$PbCl_4$	V	349.0	y	3.2	−15	d105	hyd	g	−552	−492	382	101			
chromate	$PbCrO_4$		323.2	y	6.1	844	d	i	s	−931						
fluoride	PbF_2	9, 6	245.2	w	8.2	855	1290	0.066	s	−664	−617	110	74	12		
hydroxide	$Pb(OH)_2$		241.2	w	7.6	d145		$0.016^{20°C}$	s	−516						
iodide	PbI_2	8i, 8ii	461.0	y	6.2	402	954	0.076	s	−175	−174	175	77	25	172	
molybdate	$PbMoO_4$		367.1	w	6.9	(1060)		i	s	−1052	−951	166	120			
nitrate	$Pb(NO_3)_2$		331.2	w	4.5	d470		60	s	−452						
(II) oxide	PbO	8v	223.2	y	9.5	886	d1472	0.002	s	−217	−188	69	46	26	266	15.5
(II/IV) oxide (red lead)	Pb_3O_4	9	685.6	r	9.1	d500		i	s	−718	−601	211	147			
(IV) oxide	PbO_2	7	239.2	bn	9.4	d290		i	s	−277	−217	69	65			
sulfate	$PbSO_4$		303.3	w	6.2	1170		0.0045	s	−920	−813	149	103	40		
sulfide	PbS	1	239.3	gy	7.5	1114		i	s	−100	−99	91	49	36		12.0

[e] $((PbCO_3)_2$•$Pb(OH)_2)$ – approximate formula of the precipitated carbonate.

Substance	*Formula/Structure*		M / g mol^{-1}	*col.*	ρ / g cm^{-3}	t_m / °C	t_b / °C	*sol.* / g/100 g	*State*	$\Delta_f H^\circ$ / kJ mol^{-1}	$\Delta_f G^\circ$ / kJ mol^{-1}	S° / J K^{-1} mol^{-1}	C_p° / J K^{-1} mol^{-1}	$\Delta_{fus}H$ / kJ mol^{-1}	$\Delta_{sub}H$ / kJ mol^{-1}	p / 10^{-30} C m
Lead (cont.)																
tetraethyl	$Pb(C_2H_5)_4$		323.4	nil	1.7	−137	d200	i	l	53	337	465	307		57(vap)	
tetramethyl	$Pb(CH_3)_4$		267.3	nil	2.0	−28	110	i	l	98					38(vap)	
Lithium	Li		6.94	s	0.5	180	1342	hyd	s	0	0	29	25	2.9	159	
	Li								g	159	127	139	21			
	Li^+								g	686	649	133	21			
	Li^+								aq	−278	−293	12	69			
acetate	$LiC_2H_3O_2$		66.0	w		286	d									
	$\cdot 2H_2O$		102.0	w	1.3	70	d	45								
aluminium hydride	$LiAlH_4$		37.9	w	0.9	d125		hyd	s	−116	−45	79	83			
azide	LiN_3		49.0	w	1.8	d115		70(1H_2O)	s	8						
bromide	LiBr	1	86.9	w	3.5	550	1265		s	−351	−342	74	52	18	197	24.2
	$\cdot 2H_2O$		122.9	w		d44		170	s	−963	−840	162				
carbonate	Li_2CO_3		73.9	w	2.1	723	d1310	1.29	s	−1216	−1132	90	99	45		
hydrogen	$LiHCO_3$		68.0	w				$5.5^{13°C}$	aq	−969	−881	123				
chloride	LiCl	1	42.4	w	2.1	610	1360		s	−409	−384	59	48	20	213	23.8
	$\cdot H_2O$		60.4	w	1.8	d98		85	s	−713	−632	103	98			
fluoride	LiF	1	25.9	w	2.6	845	1676	0.13	s	−616	−588	36	42	27	276	21.1
hydride	LiH	1	8.0	w	0.8	689	d950	hyd	s	−91	−68	20	28	23	230	19.6
hydroxide	LiOH		24.0	w	1.5	471	d925		s	−479	−439	43	50	21	247	15.9
	$\cdot H_2O$		42.0	w	1.5			12.9	s	−788	−681	71	80			
iodide	LiI	1	133.8	w	4.1	449	1180		s	−270	−270	87	51	15	189	24.8
	$\cdot 3H_2O$		187.9	w	3.5	d73		162	s	−1192	−1032	250	181			
nitrate	$LiNO_3$		68.9	w	2.4	261	d600		s	−483	−381	90		26		
	$\cdot 3H_2O$		123.0	w		d30		85	s	−1374	−1104	223		36		
nitride	Li_3N	9	34.8	r	1.3	813			s	−164	−128	63	76			
oxide	Li_2O	6	29.9	w	2.0	1570		hyd	s	−598	−561	38	54	59	437	
sulfate	Li_2SO_4		109.9	w	2.2	859			s	−1436	−1322	115	118	13		
	$\cdot H_2O$		128.0	w	2.1	d		35	s	−1736	−1566	164	151			

Substance	*Formula/Structure*		M g mol^{-1}	*col.*	ρ g cm^{-3}	t_m °C	t_b °C	*sol.* g/100 g	*State*	$\Delta_f H^\circ$ kJ mol^{-1}	$\Delta_f G^\circ$ kJ mol^{-1}	S° J K^{-1} mol^{-1}	C_p° J K^{-1} mol^{-1}	$\Delta_{fus}H$ kJ mol^{-1}	$\Delta_{sub}H$ kJ mol^{-1}	p 10^{-30} C m
Lithium (cont.)																
sulfide	Li_2S	6	45.9	w	1.7	900		vs	s	−441	−433	61				
Magnesium	Mg		24.31	s	1.7	650	1110	i	s	0	0	33	25	8.4	148	
	Mg								g	147	112	149	21			
	Mg^{2+}								g	2349						
	Mg^{2+}								aq	−467	−455	−137				
acetate	$Mg(C_2H_3O_2)_2$		142.4	w	1.4	d323										
	$\cdot 4H_2O$		214.4	w	1.5	80		66								
bromide	$MgBr_2$	8ii	184.1	w	3.7	711	1158		s	−524	−504	117	73	35	215	
	$\cdot 6H_2O$		292.2	w	2.0	d172		101	s	−2410	−2056	397				
carbonate	$MgCO_3$		84.3	w	3.1	d350		0.06	s	−1096	−1012	66	76			
chloride	$MgCl_2$	8i	95.2	w	2.4	714	1437		s	−641	−592	90	71	43	241	
	$\cdot 6H_2O$		203.3	w	1.6	d118		55	s	−2499	−2115	366	315	34		
fluoride	MgF_2	7	62.3	w	3.1	1248	2260	0.013	s	−1124	−1071	57	62	58	400	
hydride	MgH_2	7	26.3	w	1.4	d280		hyd	s	−75	−36	31	35			
hydroxide	$Mg(OH)_2$		58.3	w	2.4	d350		0.0012	s	−925	−834	63	77		364	
iodate	$Mg(IO_3)_2$		374.1	w	3.3	d		9.3								
	$\cdot 4H_2O$		446.2	w		d210										
iodide	MgI_2	8ii	278.1	w	4.4	d637		150(8H_2O)	s	−364	−358	130		29	192	
nitrate	$Mg(NO_3)_2$		148.3	w	1.6	89	d330	73	s	−791	−589	164	142	41		
	$\cdot 6H_2O$		256.4	w					s	−2613	−2080	452				
nitride	Mg_3N_2	9	100.9	g	2.7	d800		hyd	s	−461	−401	88	105		108	
oxide	MgO	1	40.3	w	3.6	2852	3600	i	s	−602	−569	27	37	77	617	
phosphate, hydrogen	$MgHPO_4$		120.3	w		d550										
	$\cdot 3H_2O$		174.3	w	2.1	d205		i								
phosphide	Mg_3P_2		134.9	y	2.1			hyd	s	−464	−434	77				
phosphite, hypo-	$Mg(H_2PO_2)_2$		154.3													
	$\cdot 6H_2O$		262.4	w	1.6	d100		20								

Substance	*Formula/Structure*		M g mol^{-1}	*col.*	ρ g cm^{-3}	t_m °C	t_b °C	*sol.* g/100 g	*State*	$\Delta_f H^\circ$ kJ mol^{-1}	$\Delta_f G^\circ$ kJ mol^{-1}	S° J K^{-1} mol^{-1}	C_p° J K^{-1} mol^{-1}	$\Delta_{fus}H$ kJ mol^{-1}	$\Delta_{sub}H$ kJ mol^{-1}	p 10^{-30} C m
Magnesium (cont.)																
sulfate	$MgSO_4$		120.4	w	2.7	d1124			s	−1285	−1171	92	96	15		
	$\cdot 7H_2O$		246.5	w	1.7	d150		36.4	s	−3389	−2872	372				
sulfide	MgS	1	56.4	r	2.8	d>2000		hyd	s	−346	−342	50	46		485	
Manganese	Mn		54.94	s	7.4	1244	2095	hyd	s	0	0	32	26	15	281	
	Mn								g	281	238	174	21			
	Mn^{2+}								g	2520						
	Mn^{2+}								aq	−221	−228	−74	50			
(II) acetate	$Mn(C_2H_3O_2)_2$		173.0	bn	1.7				s	−1148						
	$\cdot 4H_2O$		245.1	r	1.6	80		$43^{5\,°C}$	s	−2338						
(II) carbonate	$MnCO_3$		114.9	p	3.1	d>200		0.0065	s	−894	−817	86	82			
(II) chloride	$MnCl_2$	8i	125.8	p	3.0	650	1190		s	−481	−441	118	73	38	218	
	$\cdot 4H_2O$		197.9	p	2.0	58	d106	77	s	−1687	−1424	303				
(II) fluoride	MnF_2	7	92.9	r	4.0	856		1.05	s	−803	−761	92	67	23		
(III) fluoride	MnF_3	9	111.9	r	3.5	d>600		hyd	s	−1004	−935	105	91			
(II) hydroxide	$Mn(OH)_2$		89.0	p	3.3	d		i	s	−695	−615	99				
(II) nitrate	$Mn(NO_3)_2$		178.9						s	−576						
	$\cdot 4H_2O$		251.0	p	2.2			vs								
	$\cdot 6H_2O$		287.0	p	1.8	d26		163	s	−2372				40		
(II) oxide	MnO	1	70.9	g	5.4	1650		i	s	−385	−363	60	45	54	509	
(II/III) oxide	Mn_3O_4	9	228.8	bk	4.7	1564		i	s	−1388	−1283	156	140			
(III) oxide	Mn_2O_3	9	157.9	bk	4.5	d1080		i	s	−959	−881	110	108			
(IV) oxide	MnO_2	7	86.9	bk	5.0	d535		i	s	−520	−465	53	54			
(VII) oxide	Mn_2O_7		221.9	r	2.4	6	d55	vs	s	−728						
permanganate	MnO_4^-		118.9						aq	−541	−447	191	−82			
(II) sulfate	$MnSO_4$		151.0	p	3.2	700	d850		s	−1065	−957	112	101			
	$\cdot H_2O$		169.0	p	3.0	d400			s	−1377						
	$\cdot 5H_2O$		241.1	p	2.1	d26		65	s	−2553						
	$\cdot 7H_2O$		277.1	p	2.1	d9			s	−3139						
(II) sulfide	MnS	1, 4	87.0	p	4.0	d		i	s	−214	−218	78	50	26		

Substance	*Formula/Structure*		M g mol^{-1}	*col.*	ρ g cm^{-3}	t_m °C	t_b °C	*sol.* g/100 g	*State*	$\Delta_f H^\circ$ kJ mol^{-1}	$\Delta_f G^\circ$ kJ mol^{-1}	S° J K^{-1} mol^{-1}	C_p° J K^{-1} mol^{-1}	$\Delta_{fus}H$ kJ mol^{-1}	$\Delta_{sub}H$ kJ mol^{-1}	p 10^{-30} C m
Mercury	Hg		200.6	s	13.5	−39	357	i	l	0	0	76	28	2.3	64	0
	Hg								g	61	32	175	21			
	Hg^+								g	1075	1024	225	21			
	Hg_2^{2+}		401.2						aq	167	154	66				
	Hg^{2+}								g	2890						
	Hg^{2+}								aq	170	165	−36				
(II) acetate	$Hg(C_2H_3O_2)_2$		318.7	w	3.3	178		hyd	s	−817						
(I) bromide	Hg_2Br_2	0	561.0	w	7.3		s345	4×10^{-6}	s	−207	−181	218	105			0
(II) bromide	$HgBr_2$	0	360.4	w	6.1	236	322	0.61	s	−171	−153	172	75	18	86	0
(I) chloride	Hg_2Cl_2	0	472.1	w	7.2	*525	s383	0.0002	s	−265	−211	192	102			0
(II) chloride	$HgCl_2$	0	271.5	w	5.4	277	304	7.3	s	−224	−179	146	74	19	83	0
(II) cyanide	$Hg(CN)_2$		252.6	w	4.0	d320		11	s	264					117	
(II) fluoride	HgF_2	6	238.6	w	9.0	645	d>650	hyd	s	−423	−374	116	75	23	129	
(II) iodide	HgI_2	8v	454.4	r	6.3	259	354	0.005	s	−105	−102	180	77	19	89	0
(I) nitrate	$Hg_2(NO_3)_2$		525.2													
	•$2H_2O$		561.2	w	4.8	d70		hyd	s	−868						
(II) nitrate	$Hg(NO_3)_2$		324.6	w	4.4	79	d									
	•$\frac{1}{2}H_2O$		333.6	w	4.4	79	d	vs	s	−392						
	•H_2O		342.6	w	4.3											
(II) oxide	HgO	8a	216.6	y, r	11.1	d500		0.005	s(red)	−91	−59	70	44		133	
(I) sulfate	Hg_2SO_4		497.2	w	7.6	d		0.04	s	−743	−626	201	132			
(II) sulfate	$HgSO_4$		296.7	w	6.5	d		hyd	s	−708	−595	140				
(II) sulfide	HgS (α)	8a	232.7	r	8.1	d344		i	s(α)	−58	−51	82	48			
	HgS (β)	3	232.7	bk	7.7		s583	i	s(β)	−54	−49	88	49			
thiocyanate	$Hg(SCN)_2$		316.7	w	3.7	d165		0.063								
Molybdenum	Mo		95.96	gy	10.2	2610	5560	i	s	0	0	29	24	36	658	
	Mo								g	658	613	182	21			
(VI) fluoride	MoF_6	V	210.0	nil	2.6	17	36	hyd	l	−1586	−1473	260	170	4.1	33	0

Substance	*Formula/Structure*		M g mol^{-1}	*col.*	ρ g cm^{-3}	t_m °C	t_b °C	*sol.* g/100 g	*State*	$\Delta_f H^\circ$ kJ mol^{-1}	$\Delta_f G^\circ$ kJ mol^{-1}	S° J K^{-1} mol^{-1}	C_p° J K^{-1} mol^{-1}	$\Delta_{fus}H$ kJ mol^{-1}	$\Delta_{sub}H$ kJ mol^{-1}	p 10^{-30} C m
Molybdenum (cont.)																
(IV) oxide	MoO_2	7	128.0	gy	6.5	d1100		i	s	−589	−533	46	56		602	
(VI) oxide	MoO_3	8v	144.0	w	4.7	801	1107		s	−745	−668	78	75	49	419	
(molybdic acid)	$\bullet 2H_2O$		180.0	y	3.1	d70		0.21	s	−1360						
(III) sulfide	Mo_2S_3	9	288.1	gy	5.9	1807	d1867		s	−407	−396	115	109	130		
(IV) sulfide	MoS_2	8v	160.1	bk	4.8	1185		i	s	−235	−226	63	64			
Nickel	Ni		58.69	s	8.9	1455	2730	i	s	0	0	30	26	18	430	
	Ni								g	430	385	182	23			
	Ni^{2+}								g	2931						
	Ni^{2+}								aq	−54	−46	−129				
acetate	$Ni(C_2H_3O_2)_2$		176.8	g	1.8	d										
	$\bullet 4H_2O$		248.8	g	1.7	d		16								
arsenide	NiAs	5	133.6	r	7.8	968		i								
bromide	$NiBr_2$	8ii	218.5	y	5.1	963		134($6H_2O$)	s	−212	−198	136		54	246	
carbonate	$NiCO_3$		118.7	g	4.4	d		0.009	s	−681	−613	118				
hydroxide	$NiCO_3 \bullet 2Ni(OH)_2 \bullet 4H_2O$[f]		376.2	g	2.6			i								
carbonyl	$Ni(CO)_4$		170.7	w	1.3	−25	43	0.02	l	−633	−588	313	205	14	44	0
chloride	$NiCl_2$	8i	129.6	y	3.6	*1000	s973		s	−305	−259	98	72	77	247	
	$\bullet 6H_2O$		237.7	g				66	s	−2103	−1714	344				
fluoride	NiF_2	7	96.7	g	4.7	1450	1740	2.6($4H_2O$)	s	−651	−604	74	64		273	
hydroxide	$Ni(OH)_2$		92.7	g	4.1	d230		0.001	s	−530	−447	88				
iodide	NiI_2	8i	312.5	bk	5.8	797		154	s	−78	−81	154				
nitrate	$Ni(NO_3)_2$		182.7						s	−415						
	$\bullet 6H_2O$		290.8	g	2.1	57	137	100	s	−2212			464			
oxide	NiO	1	74.7	gy	6.7	1984		i	s	−240	−212	38	44		554	
sulfate	$NiSO_4$		154.8	y	3.7	d840			s	−873	−760	92	138			
	$\bullet 6H_2O$		262.8	g	2.1	d53			s	−2683	−2225	334	328			
	$\bullet 7H_2O$		280.9	g	1.9	d98		41	s	−2976	−2462	379	365			
sulfide	NiS	5, 9	90.8	bk	5.3	976		i	s	−82	−80	53	47	30		

[f] ($NiCO_3 \bullet 2Ni(OH)_2 \bullet 4H_2O$) – approximate formula of the precipitated carbonate.

Substance	*Formula/Structure*		M g mol^{-1}	*col.*	ρ g cm^{-3}	t_m °C	t_b °C	*sol.* g/100 g	*State*	$\Delta_f H^\circ$ kJ mol^{-1}	$\Delta_f G^\circ$ kJ mol^{-1}	S° J K^{-1} mol^{-1}	C_p° J K^{-1} mol^{-1}	$\Delta_{fus}H$ kJ mol^{-1}	$\Delta_{sub}H$ kJ mol^{-1}	p 10^{-30} C m
Nitrogen	N_2		28.02	nil	0.8(l)	−210	−196	0.0017	g	0	0	192	29	0.72		0
	N		14.01						g	473	456	153	21			
	N_3^-		42.0						g	181						
	N_3^-		42.0						aq	275	348	108				
acid, hydrazoic	HN_3	V	43.0	nil	1.1(l)	−80	37	∞	l	264	327	141			30(vap)	5.7
acid, nitric	HNO_3 fuming		63.0	nil	1.5(l)	−42	83	∞	l	−174	−81	156	110	10	49	7.2
	NO_3^-		62.0						aq	−207	−111	147	−87			
ammonia[g]	NH_3	0	17.0	nil	0.8(l)	−78	−33	46	g	−46	−16	193	35	5.7	26	4.9
chloride, tri-	NCl_3	0	120.4	y	1.7	−27	71	i	l	230						1.3
cyanamide	NH_2CN		42.0	w	1.3	42		s	s	59						
fluoride, tri-	NF_3	0	71.0	nil	1.5(l)	−207	−129	i	g	−125	−83	261	53			0.8
hydrazine	N_2H_4	0	32.0	nil	1.0	1	114	vs	l	51	149	121	99	13	58	5.8
hydrate	$\cdot H_2O$		50.1	nil	1.0	−40	119	∞	g	−205	−79	264				
hydrochloride	$N_2H_4\cdot HCl$		68.5	w	1.5	89	d240	vs	s	−197				15		
	$N_2H_4\cdot 2HCl$		105.0	w	1.4	d198		$27^{32°C}$	s	−367						
sulfate	$N_2H_4\cdot H_2SO_4$		130.1	w	1.4	254	d	3.4	s	−959						
hydrazine, 1,1-dimethyl	$(CH_3)_2NNH_2$		60.1		0.8	−57	64	vs	l	49	207	198	164		35(vap)	
hydrazine, phenyl	$C_6H_5NHNH_2$		108.1	r-bn	1.1	21	243	s	l	141					62(vap)	
hydroxylamine	NH_2OH		33.0	w	1.2	33	57	s	s	−114						2.0
hydrochloride	$NH_2OH\cdot HCl$		69.5	w	1.7	151	d	94	s	−318			93			
sulfate	$(NH_2OH)_2\cdot H_2SO_4$		164.1	w		(170)		s								
(di-) oxide	N_2O	0	44.0	nil	1.2(l)	−91	−88	0.112	g	82	104	220	38	6.5		0.6
oxide	NO	0(d)	30.0	nil	1.3(l)	−164	−152	0.0056	g	90	87	211	30	2.3		0.5
(di-) oxide, tri-	N_2O_3	–	76.0	r, b	1.4(l)	−102	d3	s	g	84	139	312	66		33(vap)	7.1
(di-) oxide, tetra-	N_2O_4	0	92.0	nil	1.4(l)	−11	21	hyd	g	9	98	304	77	15	44	0
oxide, di-	NO_2		46.0						g	33	51	240	37			1.0
(di-) oxide, penta-	N_2O_5	9	108.0	w	2.0	30	d47	s	g	11	115	356	84		54	
									aq	−105	−32	123	−97			

[g] See Ammonia and Ammonium compounds, pages 37–9.

Substance	*Formula/Structure*		M / g mol^{-1}	*col.*	ρ / g cm^{-3}	t_m / °C	t_b / °C	*sol.* / g/100 g	*State*	$\Delta_f H^\circ$ / kJ mol^{-1}	$\Delta_f G^\circ$ / kJ mol^{-1}	S° / J K^{-1} mol^{-1}	C_p° / J K^{-1} mol^{-1}	$\Delta_{fus}H$ / kJ mol^{-1}	$\Delta_{sub}H$ / kJ mol^{-1}	p / 10^{-30} C m
Nitrogen (cont.)																
nitrosyl chloride	NOCl		65.5	o	1.4(l)	−62	−6	hyd	g	52	66	262	45			6.3
sulfide	N_4S_4	0	184.3	o	2.2	d178		i	s	536						
Osmium	Os		190.2	b-w	22.6	3033	5012	i	s	0	0	33	25	58	791	
	Os								g	791	745	193	21			
(VIII) oxide	OsO_4	0	254.2	y	5.1	41	131	7.0	s	−394	−305	144	74(g)	14	57	0
Oxygen	O_2		32.00	nil	1.1(l)	−219	−183	0.004	g	0	0	205	29	0.44		0
	O								g	249	232	161	22			
	O^-								g	102	92	158	22			
fluoride, di-	OF_2	V	54.0	nil	1.9(l)	−224	−145	hyd	g	25	42	247	43			1.0
ozone	O_3		48.0	nil	1.6(l)	−193	−111	0.0005	g	143	163	239	39			1.8
Palladium	Pd		106.4	s	12.0	1554	2970	i	s	0	0	38	26	17		
	Pd			gy					g	378	340	167	21			
	Pd^{2+}								g	3069						
	Pd^{2+}								aq	149	176	−184				
acetate	$Pd(C_2H_3O_2)_2$		224.5	o-bn		d205		i								
acetylacetonate	$Pd(C_5H_7O_2)_2$		304.6	y		d>190		i								
chloride	$PdCl_2$		177.3	r	4.0	d500		670	s	−199						
nitrate	$Pd(NO_3)_2$		230.4	y		d		hyd								
	$\cdot 2H_2O$		266.5													
oxide	PdO	9	122.4	g-bk	8.3	d750		i	s	−85			31			
sulfate	$PdSO_4$		202.5													
	$\cdot 2H_2O$		238.5	g-bn		d>650		s								
Phosphorus	P (white)		30.97	w	1.8	44	280	i	s(white)	0	0	41	24	0.66	15	
	P (red)		30.97	r	2.3	590	s417	i	s(red)	−18	−12	23	21	19	32	
	P (black)		30.97	bk	2.7			i	s(black)	−13	−7	23	22			
	P								g	317	278	163	21			
	P_4		123.9						g	59	24	280	67		59	
acid, phosphoric	H_3PO_4		98.0	w	1.8	42	d213	612	s	−1279	−1119	110	106	13		
phosphate	PO_4^{3-}		95.0						aq	−1277	−1019	−220				

Substance	*Formula/Structure*	M g mol^{-1}	*col.*	ρ g cm^{-3}	t_m °C	t_b °C	*sol.* g/100 g	*State*	$\Delta_f H^\circ$ kJ mol^{-1}	$\Delta_f G^\circ$ kJ mol^{-1}	S° J K^{-1} mol^{-1}	C_p° J K^{-1} mol^{-1}	$\Delta_{fus}H$ kJ mol^{-1}	$\Delta_{sub}H$ kJ mol^{-1}	p 10^{-30} C m
Phosphorus (cont.)															
acid, phosphorous	H_3PO_3	82.0	w	1.6	74	d200	vs	s	−964				13		
acid, hypophos-phorous	H_3PO_2	66.0	w	1.5	27	d130	s	s	−605				9.7		
bromide, tri-	PBr_3 V	270.7	nil	2.8	−40	173	hyd	l	−184	−176	240			45(vap)	
bromide, penta-	PBr_5 9	430.5	y	3.5	d106		hyd	s	−270					60	
bromide, oxy-	$POBr_3$	286.7	w	2.8	56	d192	hyd	s	−459					66	
bromide, thio-	$PSBr_3$	302.8	y	2.8	38	d212	s	s			231				
chloride, tri-	PCl_3 V	137.3	nil	1.6	−94	76	hyd	l	−320	−272	217		4.5	37	1.9
chloride, penta-	PCl_5 9	208.2	w	2.1	d167	s160	hyd	g	−375	−305	365	113		69	0
chloride, oxy-	$POCl_3$	153.3	nil	1.7	1	105	hyd	l	−597	−521	222	139	13	52	8.5
fluoride, tri-	PF_3 V	88.0	nil	4.0(l)	−151	−101	hyd	g	−919	−898	273	59			3.4
fluoride, penta-	PF_5 V	126.0	nil		−94	−85	hyd	g	−1594	−1521	301	85	12		0
hydride, tri-, (phosphine)	PH_3 V	34.0	nil	0.7(l)	−133	−88	vs	g	5	13	210	37	1		1.9
phosphine, triphenyl	$P(C_6H_5)_3$	262.3	w-y	1.2	79		i	s	232					96	
phosphonium															
bromide	PH_4Br	114.9	w			s30	hyd	s	−128	−48	110				
chloride	PH_4Cl	70.5	w		*28	s	hyd	s	−145						
iodide	PH_4I	161.9	nil	2.9	19	80	hyd	s	−70	1	123	110			
(di-)hydride, tetra-	P_2H_4 V	66.0	nil	1.0	−99	67	hyd	l	−5					26(vap)	
iodide, tri-	PI_3 V	411.7	o	4.2	61	d	hyd	s	−46					28	
nitride	PN V	45.0	r		d750			g	110	88	211	30		173	9.2
(tetra-)oxide, hexa-	P_4O_6 V	219.9	w	2.1	24	173	hyd	l	−1646				19	67	
oxide, di-	PO_2 –	63.0	w	2.5	>100	180	hyd	g	−280	−282	252	40			
(di-)oxide, penta-	P_2O_5 0², 9, 8v	141.9	w	2.4	*580	s300	hyd	s	−2984	−2700	229	212	48	106	
(di-)sulfide, tri-	P_2S_3 –	158.1	y		290	490	hyd	s	−80						
(di-)sulfide, penta-	P_2S_5 0(d)	222.3	y	2.0	286	514	i	s	−186	−171	191	148	21		

Substance	*Formula/Structure*		M / g mol^{-1}	*col.*	ρ / g cm^{-3}	t_m / °C	t_b / °C	*sol.* / g/100 g	*State*	$\Delta_f H^\circ$ / kJ mol^{-1}	$\Delta_f G^\circ$ / kJ mol^{-1}	S° / J K^{-1} mol^{-1}	C_p° / J K^{-1} mol^{-1}	$\Delta_{fus}H$ / kJ mol^{-1}	$\Delta_{sub}H$ / kJ mol^{-1}	p / 10^{-30} C m
Platinum	Pt		195.1	s-gy	21.4	1772	3825	i	s	0	0	42	26	22	565	
	Pt								g	565	520	192	26			
acetylacetonate	$Pt(C_5H_7O_2)_2$		393.3	y		250	d420	i								
(II) chloride	$PtCl_2$		266.0	g	6.0	d581		i	s	−123						
(IV) chloride	$PtCl_4$	8a	336.9	bn	4.3	d327		142	s	−232	−148	205				
(II) oxide	PtO		211.1	bk	14.1	d325		i								
(IV) oxide	PtO_2	7	227.1	bk	11.8	450		i	g	172	168	259				
	$\bullet xH_2O$			bk		d>100		i								
Potassium	K		39.10	s	0.9	63	760	hyd	s	0	0	65	30	2.4	89	0.0
	K								g	89	61	160	21			
	K^+								g	514	481	155	21			
	K^+								aq	−252	−284	101	22			
acetate	$KC_2H_3O_2$		98.1	w	1.6	292		270($1\frac{1}{2}H_2O$)	s	−723						
aluminium sulfate	See Aluminium, page 16															
arsenate	K_3AsO_4		256.2	w	2.8	1310		125($7H_2O$)								
hydrogen	K_2HAsO_4		218.1	w		d300		175($3H_2O$)								
dihydrogen	KH_2AsO_4		180.0	w	2.9	288		33($1H_2O$)	s	−1181	−1036	155	127			
arsenite, meta-	$KAsO_2$		146.0	w				s								
arsenite, ortho-	K_3AsO_3		240.2	w				vs								
borate, tetra-	$K_2B_4O_7$		233.4	w	1.7											
	$\bullet 4H_2O$		305.5	w		d112		$21^{35°C}$	s		−4135					
tetrafluoro-	KBF_4		125.9	w	2.5	530		0.57	s	−1882	−1786	152	111	18	335	
bromate	$KBrO_3$		167.0	w	3.3	*434	d370	8.1	s	−360	−271	149	105			
bromide	KBr	1	119.0	w	2.7	734	1435	68	s	−394	−381	96	52	26	214	34.7
carbonate	K_2CO_3		138.2	w	2.4	891	d		s	−1151	−1064	156	114	28		
	$\bullet 1\frac{1}{2}H_2O$		165.2	w	2.0			112	s	−1609	−1432	203				
	$\bullet 2H_2O$		174.2	w	2.0	d130										
hydrogen	$KHCO_3$		100.1	w	2.2	d100		36	s	−963	−864	116				
chlorate	$KClO_3$		122.6	w	2.3	356	d400	8.6	s	−398	−296	143	100			

Substance	*Formula/Structure*		M g mol^{-1}	*col.*	ρ g cm^{-3}	t_m °C	t_b °C	*sol.* g/100 g	*State*	$\Delta_f H^\circ$ kJ mol^{-1}	$\Delta_f G^\circ$ kJ mol^{-1}	S° J K^{-1} mol^{-1}	C_p° J K^{-1} mol^{-1}	$\Delta_{fus}H$ kJ mol^{-1}	$\Delta_{sub}H$ kJ mol^{-1}	p 10^{-30} C m
Potassium (cont.)																
chlorate, per-	$KClO_4$		138.6	w	2.5	d400		2.1	s	−433	−303	151	112			
chloride	KCl	1	74.6	w	2.0	770	1437	36	s	−437	−409	83	51	27	223	34.3
chromate	K_2CrO_4		194.2	y	2.7	975		65	s	−1404	−1296	200	146	29		
chromate, di-	$K_2Cr_2O_7$		294.2	o	2.7	398	d500	15	s	−2062	−1882	291	219	36		
chromium sulfate	See Chromium, page 48															
citrate	$K_3C_6H_5O_7$		306.4	w		d230										
	$\cdot H_2O$		324.4	w	2.0	d180		172								
cyanate	KCNO		81.1	w	2.1	d700		75	s	−419						
cyanide	KCN		65.1	w	1.5	622	1625	72	s	−113	−102	128	66	15	204	
ferricyanide	$K_3Fe(CN)_6$		329.2	o	1.9	d		49	s	−250	−130	426				
ferrocyanide	$K_4Fe(CN)_6$		368.3			d			s	−594	−453	419	332			
	$\cdot 3H_2O$		422.4	y	1.9	d70		32	s	−1467	−1169	594	482			
fluoride	KF	1	58.1	w	2.5	858	1505		s	−567	−538	67	49	28	242	28.7
	$\cdot 2H_2O$		94.1	w	2.5	41		102	s	−1164	−1021	155				
formate	$KCHO_2$		84.1	w	1.9	168	d	$331^{18\,°C}$	s	−680						
hydride	KH	1	40.1	w	1.4	d417		hyd	s	−58	−53	50	38		181	
hydroxide	KOH		56.1	w	2.0	406	1320	$119(2H_2O)$	s	−425	−379	79	65	9.4	194	
iodate	KIO_3		214.0	w	3.9	d560		9.2	s	−501	−418	151	106			
iodate, per-	KIO_4		230.0	w	3.6	582	d	0.51	s	−467	−361	176				
iodide	KI	1	166.0	w	3.1	681	1330	148	s	−328	−325	106	53	24	202	
nitrate	KNO_3		101.1	w	2.1	334	d400	38	s	−495	−395	133	96	12		
nitrite	KNO_2		85.1	w	1.9	*440	d350	312	s	−370	−307	152	107			
oxalate	$K_2C_2O_4$		166.2	w					s	−1347						
	$\cdot H_2O$		184.2	w	2.1	d160		38	s	−1645						
oxide, mono-	K_2O	6	94.2	w	2.3	d350		hyd	s	−363	−322	94	84			
oxide, per-	K_2O_2	–	110.2	w		490	d	hyd	s	−494	−425	102	100			
oxide, super-	KO_2	9	71.1	y	2.1	380	d	hyd	s	−285	−239	117	78			

Substance	*Formula/Structure*	M g mol^{-1}	*col.*	ρ g cm^{-3}	t_m °C	t_b °C	*sol.* g/100 g	*State*	$\Delta_f H^\circ$ kJ mol^{-1}	$\Delta_f G^\circ$ kJ mol^{-1}	S° J K^{-1} mol^{-1}	C_p° J K^{-1} mol^{-1}	$\Delta_{fus} H$ kJ mol^{-1}	$\Delta_{sub} H$ kJ mol^{-1}	p 10^{-30} C m
Potassium (cont.)															
permanganate	$KMnO_4$	158.0	v	2.7	d240		7.6	s	−837	−738	172	118			
phosphate	K_3PO_4	212.3	w	2.6	1340		106(7H_2O)	s	−1950				37		
	•$3H_2O$	266.3													
dihydrogen	KH_2PO_4	136.1	w	2.3	253		25	s	−1568	−1416	135	117			
pyro-	$K_4P_2O_7$	330.3													
	•$3H_2O$	384.4	w	2.3	d180		s								
phosphite															
hydrogen	K_2HPO_3	158.2	w		d		vs								
dihydrogen	KH_2PO_3	120.1	w		d		$172^{20°C}$								
hypo-	KH_2PO_2	104.1	w		d		200								
phthalate															
hydrogen	$KHC_8H_4O_4$	204.2	w	1.6	d		11								
selenate	K_2SeO_4	221.2	w	3.1			110	s	−1110	−1003	222				
selenide	K_2Se 6	157.2	w	2.3	800		hyd	s	−395						
selenite	K_2SeO_3	205.2	w		d875		217	s	−982						
silicate, hexafluoro-	K_2SiF_6	220.3	w	2.3	d		0.15	s	−2956	−2799	226				
sulfate	K_2SO_4	174.3	w	2.7	1069	1689	12.0	s	−1438	−1321	176	131	37		
hydrogen	$KHSO_4$	136.2	w	2.3	214	d	52	s	−1161	−1031	138				
peroxydi-	$K_2S_2O_8$	270.3	w	2.5	d100		6.2	s	−1916	−1697	279	213			
pyro-	$K_2S_2O_7$	254.3	w	2.3	>300	d	s	s	−1987	−1792	255				
sulfide	K_2S 6	110.3	bn	1.8	948		s	s	−376	−363	115	75	16	82	
	•$5H_2O$	200.3	w		60	d150	s	s	−1872						
sulfite	K_2SO_3	158.2	w				100	s	−1125						
	•$2H_2O$	194.2	w		d										
pyro-	$K_2S_2O_5$	222.3	w	2.3	d190		49	s	−1534						
tartrate	$K_2C_4H_4O_6$	226.3			d200										
	•$\frac{1}{2}H_2O$	235.3	w	2.0	d155		$140^{17°C}$								
hydrogen	$KHC_4H_4O_6$	188.2	w	2.0			0.6								
sodium	$KNaC_4H_4O_6$	210.1	w												
	•$4H_2O$	282.2	w	1.8	70	d130	$90^{20°C}$								

Substance	*Formula/Structure*		M g mol^{-1}	*col.*	ρ g cm^{-3}	t_m °C	t_b °C	*sol.* g/100 g	*State*	$\Delta_f H^\circ$ kJ mol^{-1}	$\Delta_f G^\circ$ kJ mol^{-1}	S° J K^{-1} mol^{-1}	C_p° J K^{-1} mol^{-1}	$\Delta_{fus}H$ kJ mol^{-1}	$\Delta_{sub}H$ kJ mol^{-1}	p 10^{-30} C m
Potassium (cont.)																
tellurate																
tetrahydrogen	$K_2H_4TeO_6$		305.8			d300										
	$\bullet 3H_2O$		359.9	w		d		i								
meta-	K_2TeO_4		269.8			d200		hyd								
telluride	K_2Te	6	205.8	w	2.5			hyd								
tellurite	K_2TeO_3		253.8	w		d460		vs								
thiocyanate	KSCN		97.2	w	1.9	173	d500	239	s	−200	−178	124	89	10		
Rhenium	Re		186.2	s-gy	21.0	3185	5596	i	s	0	0	37	25	34	770	
	Re								g	770	725	189	21			
(VII) oxide	Re_2O_7	0	484.4	y	6.1	327	360	s	s	−1240	−1066	207	166	66		
Rhodium	Rh		102.9	s-w	12.4	1964	3695	i	s	0	0	32	25	27	557	
	Rh								g	557	511	186	21			
	Rh^{3+}								g	6042						
(III) chloride	$RhCl_3$		209.3	r	5.4		717	i	s	−299	−228	127				
	$\bullet 3H_2O$		263.3	r-bn												
(III) nitrate	$Rh(NO_3)_3$		288.9	bn		d600		i								
(III) oxide	Rh_2O_3	9	253.8	gy	8.2	d1100		i	s	−343	−260	92	104			
Rubidium	Rb		85.47	s	1.5	39	686	hyd	s	0	0	77	31	2.2	81	
	Rb								g	81	53	170	21			
	Rb^+								g	490						
	Rb^+								aq	−251	−284	122				
bromide	RbBr	1	165.4	w	3.4	682	1340	116	s	−395	−382	110	53	15	212	
carbonate	Rb_2CO_3		230.9	w		*837	d740	450	s	−1179	−1096	186	119			
chloride	RbCl	1	120.9	w	2.8	715	1390	94	s	−435	−408	96	52	18	207	35.0
fluoride	RbF	1	104.5	w	3.6	795	1410	301 (18°C)	s	−558	−521	75	51	23	226	28.5
hydride	RbH	1	86.5	w	2.6	d300		hyd	s	−52						
hydroxide	RbOH		102.5	w	3.2	301		178	s	−418				7	180	
iodide	RbI	1	212.4	w	3.6	647	1300	163	s	−334	−329	118	53	13	200	

Substance	*Formula/Structure*		*M* g mol⁻¹	*col.*	*ρ* g cm⁻³	t_m °C	t_b °C	*sol.* g/100 g	*State*	$\Delta_f H^\circ$ kJ mol⁻¹	$\Delta_f G^\circ$ kJ mol⁻¹	S° J K⁻¹ mol⁻¹	C_p° J K⁻¹ mol⁻¹	$\Delta_{fus} H$ kJ mol⁻¹	$\Delta_{sub} H$ kJ mol⁻¹	*p* 10^{-30} C m
Rubidium (cont.)																
nitrate	$RbNO_3$		147.5	w	3.1	310	d	65	s	−495	−396	147	102	5.6		
oxide	Rb_2O	6	186.9	y	3.7	d400		hyd	s	−339					289	
oxide, per-	Rb_2O_2	9	202.9	y	3.6	570	d1011	hyd	s	−472						
oxide, super-	RbO_2	9	117.5	y	3.8	432	d1157	hyd	s	−279						
sulfate	Rb_2SO_4		267.0	w	3.6	1060	(1700)	51	s	−1436	−1317	197	134		367	
sulfide	Rb_2S	6	203.0	y	2.9	d530		vs	s	−361	−324	133				
Ruthenium	Ru		101.1	s-w	12.1	2334	4150	i	s	0	0	29	24	39		
	Ru								g	643	596	186	22			
(III) chloride	$RuCl_3$	8v	207.4	bn	3.1	d>500			s	−205	−135	128				
	$\cdot xH_2O$			bn	3.1	d100		s								
acetylacetonate	$Ru(C_5H_7O_2)_3$		398.4	r		230	d>250	i								
(IV) oxide	RuO_2		133.1	gy-bk	7.0	d1300		i	s	−305	−253	61				
	$\cdot xH_2O$			bk		d>75		i								
Scandium	Sc		44.96	s	3.0	1538	2730	hyd	s	0	0	35	26	16	378	
	Sc								g	378	336	175	22			
	Sc^{3+}								g	4652						
	Sc^{3+}								aq	−614	−587	−255				
chloride	$ScCl_3$	8iv	151.3	w	2.4	939		$76^{0°C}$	s	−925	−851	121	92	67		
hydroxide	$Sc(OH)_3$		96.0	w				i	s	−1364	−1233	100				
nitride	ScN	1	59.0	b		2650		i	s	−314	−284	30	37			
oxide	Sc_2O_3	9	137.9	w	3.9	>2400		i	s	−1909	−1819	77	94			
Selenium	Se (grey)		78.96	gy	4.8	217	685	i	s(grey)	0	0	42	25	5.1	227	
	Se (red)		78.96	r	4.5	d170		i	s(red)	7					220	
	Se								g	227	187	177	21			
	Se_2		157.9						g	146	96	252	35			
acid, selenic	H_2SeO_4		145.0	w	3.0	58	d260	1060	s	−530				14		
selenate	SeO_4^{2-}		143.0						aq	−599	−441	54				

Substance	Formula/Structure	M g mol^{-1}	col.	ρ g cm^{-3}	t_m °C	t_b °C	sol. g/100 g	State	$\Delta_f H^\circ$ kJ mol^{-1}	$\Delta_f G^\circ$ kJ mol^{-1}	S° J K^{-1} mol^{-1}	C_p° J K^{-1} mol^{-1}	$\Delta_{fus}H$ kJ mol^{-1}	$\Delta_{sub}H$ kJ mol^{-1}	p 10^{-30} C m
Selenium (cont.)															
acid, selenous	H_2SeO_3	129.0	w	3.0	d70		203	s	−524	−370	13				
selenite	SeO_3^{2-}	127.0						aq	−509						
chloride, tetra-	$SeCl_4$ 9	220.8	y	3.8	*305	s196	hyd	s	−183	−95	195				
fluoride, hexa-	SeF_6 9	193.0	nil		*−39	s−47	hyd	g	−1117	−1017	314	110	8.4	27	0
oxide, di-	SeO_2 8a	111.0	w	4.0	*340	s317	274	s	−225	−171	67	58	29	117	8.8
oxide, tri-	SeO_3 0(t), 8a	127.0	y	3.6	118	d180	hyd	s	−167						
Silicon	Si	28.09	gy	2.3	1410	3267	i	s	0	0	19	20	50		
	Si							g	450	406	168	22			
bromide, tetra-	$SiBr_4$ –	347.7	w	2.8	5	154	hyd	l	−457	−444	278	146		42(vap)	0
carbide	SiC 3, 4	40.1	bk	3.2	d2986		i	s	−65	−63	17	27		806	
chloride, tetra-	$SiCl_4$ V	169.9	nil	1.5(l)	−70	57	hyd	l	−687	−620	240	145	7.7	38	0
fluoride, tetra-	SiF_4 0	104.1	nil	1.7(l)	*−90	s−96	hyd	g	−1615	−1573	283	74		26	0
hydride (silane)	SiH_4 V	32.1	nil	0.7(l)	−185	−112	i	g	34	57	205	43	0.67		0
(disilane)	Si_2H_6 V	62.2	nil	0.7(l)	−133	−15	hyd	g	80	127	273	81			0
(trisilane)	Si_3H_8 V	92.3	nil	0.7(l)	−117	53	hyd	l	92					28(vap)	
(tetrasilane)	Si_4H_{10} V	122.4	nil	0.8	−84	107	hyd								
iodide, tetra-	SiI_4 0	535.7	w	4.2	120	288	hyd	s	−199	−203	265	108	15	77	0
nitride	Si_3N_4 9	140.3	gy	3.4	*1900	d1878	i	s	−743	−643	101	100			
oxide, di-	SiO_2 9	60.1	w	2.6	1713	2230	0.012	s(quartz)	−911	−856	41	44	11	589	
sulfide, di-	SiS_2 8a	92.2	w	2.0		s1090	hyd	s	−213	−213	80	77	8		
Silver	Ag	107.9	s	10.5	962	2212	i	s	0	0	43	25	11	285	
	Ag							g	285	246	173	21			
	Ag^+							g	1022						
	Ag^+							aq	106	77	73	22			
acetate	$AgC_2H_3O_2$	166.9	w	3.3	d		1.11	s	−399	−308	150				
bromide	AgBr 1	187.8	y	6.5	430	d>1300	i	s	−100	−97	107	52	9.2		
carbonate	Ag_2CO_3	275.8	y	6.1	d218		0.003	s	−506	−437	167	112			

Substance	*Formula/Structure*	M g mol^{-1}	*col.*	ρ g cm^{-3}	t_m °C	t_b °C	*sol.* g/100 g	*State*	$\Delta_f H^\circ$ kJ mol^{-1}	$\Delta_f G^\circ$ kJ mol^{-1}	S° J K^{-1} mol^{-1}	C_p° J K^{-1} mol^{-1}	$\Delta_{fus}H$ kJ mol^{-1}	$\Delta_{sub}H$ kJ mol^{-1}	p 10^{-30} C m
Silver (cont.)															
chlorate, per-	$AgClO_4$	207.3	w	2.8	d486		557(1H_2O)	s	−31						
chloride	$AgCl$ 1	143.3	w	5.6	455	1564	0.00019	s	−127	−110	96	51	13	224	19.0
chromate	Ag_2CrO_4	331.7	r	5.6			0.0025	s	−732	−642	218	142			
cyanate	$AgCNO$	149.9	w	4.0	d		$0.007^{18°C}$	s	−95	−58	121				
cyanide	$AgCN$	133.9	w	3.9	d320		0.007	s	146	157	107	67	12		
fluoride	AgF 1	126.9	y	5.9	435	(1159)	180(2H_2O)	s	−205	−187	84				20.8
iodide	AgI 4, 3, –, 1	234.8	y	5.7	558	1506	i	s	−62	−66	115	57	9.4		17.0
nitrate	$AgNO_3$	169.9	w	4.4	212	d440	245	s	−124	−33	141	93	12		
nitrite	$AgNO_2$	153.9	w	4.5	d140		0.41	s	−45	19	128	80			
oxalate	$Ag_2C_2O_4$	303.8	w	5.0	d140		0.0041	s	−673	−584	209				
oxide	Ag_2O 9	231.7	bn	7.1	d200		0.0022	s	−31	−11	121	66			
phosphate	Ag_3PO_4	418.6	y	6.4	849		0.0006	s		−879					
sulfate	Ag_2SO_4	311.8	w	5.5	652	d1085	0.83	s	−716	−618	200	131	18		
sulfide	Ag_2S –, 9	247.8	bk	7.3	825	d	i	s	−33	−41	144	77	14		
thiocyanate	$AgSCN$	166.0	w		d		1×10^{-5}	s	88	101	131	63			
Sodium	Na	22.99	s	1.0	98	883	hyd	s	0	0	51	28	2.6	107	0
	Na							g	107	77	154	21			
	Na^+							g	609	574	148	21			
	Na^+							aq	−240	−262	58	46			
acetate	$NaC_2H_3O_2$	82.0	w	1.5	324			s	−709	−607	123	80			
	•3H_2O	136.1	w	1.4	d58		51	s	−1603	−1329	243				
aluminate	$NaAlO_2$	82.0	w	4.6	1800		vs	s	−1135	−1071	71	73			
ammonium															
phosphate															
hydrogen	$NaNH_4HPO_4$	137.0													
	•4H_2O	209.1	w	1.6	d79		17	s	−2852						

Substance	*Formula/Structure*	M g mol^{-1}	*col.*	ρ g cm^{-3}	t_m °C	t_b °C	*sol.* g/100 g	*State*	$\Delta_f H^\circ$ kJ mol^{-1}	$\Delta_f G^\circ$ kJ mol^{-1}	S° J K^{-1} mol^{-1}	C_p° J K^{-1} mol^{-1}	$\Delta_{fus}H$ kJ mol^{-1}	$\Delta_{sub}H$ kJ mol^{-1}	p 10^{-30} C m
Sodium (cont.)															
arsenate	Na_3AsO_4	207.9						s	−1540						
	•$12H_2O$	424.1	w	1.8	86		$12^{17°C}$	s	−5092						
hydrogen	Na_2HAsO_4	185.9			d150										
	•$7H_2O$	312.0	w	1.9	d50		41.4								
dihydrogen	NaH_2AsO_4	163.9			d200										
	•H_2O	181.9	w	2.5	d100		s								
arsenite	$NaAsO_2$	129.9	w	1.9			s	s	−661						
azide	NaN_3	65.0	w	1.8	d		$41^{20°C}$	s	22	94	97	77			
benzoate	$NaC_7H_5O_2$	144.1	w				63								
bismuthate	$NaBiO_3$	280.0	y		d		i								
borate, meta-	$NaBO_2$	65.8	w	2.5	966	1434		s	−977	−921	74	66	36	333	
	•$4H_2O$	137.9	w		57	d120	28	s	−2176	−1888	230				
borate, per-	$NaBO_3$	81.8													
	•$4H_2O$	153.9	w	1.7	d60		hyd	s	−2114						
borate, tetra-	$Na_2B_4O_7$	201.2	w	2.4	741	d1575		s	−3291	−3096	190	187	81		
(borax)	•$10H_2O$	381.4	w	1.7	d75		3.2	s	−6289	−5516	586	615			
borohydride	$NaBH_4$	37.8	w	1.1	d400		55	s	−189	−124	101	87			
bromate	$NaBrO_3$	150.9	w	3.3	381	d	39	s	−334	−243	129		28		
bromide	$NaBr$ 1	102.9	w	3.2	747	1390		s	−361	−349	87	51	26	218	30.4
	•$2H_2O$	138.9	w	2.2	d51		95	s	−952	−828	179				
carbonate	Na_2CO_3	106.0	w	2.5	851	d		s	−1131	−1044	135	112	29		
	•H_2O	124.0	w	2.3	d100			s	−1431	−1285	168	146			
	•$7H_2O$	232.1	w	1.5	d32			s	−3200	−2714	422				
	•$10H_2O$	286.1	w	1.4	d34		29.4	s	−4081	−3428	563	550			
hydrogen	$NaHCO_3$	84.0	w	2.2	d270		10.3	s	−951	−851	102	88			
chlorate	$NaClO_3$	106.4	w	2.5	248	d>300	100	s	−366	−262	123		23		
chlorate, per-	$NaClO_4$	122.4	w	2.5	d482			s	−383	−255	142	111			
	•H_2O	140.5	w	2.0	d130		210	s	−678	−494	191				
chloride	$NaCl$ 1	58.4	w	2.2	801	1465	35.9	s	−411	−384	72	50	28	235	30.0

Substance	*Formula/Structure*	M g mol^{-1}	*col.*	ρ g cm^{-3}	t_m °C	t_b °C	*sol.* g/100 g	*State*	$\Delta_f H^\circ$ kJ mol^{-1}	$\Delta_f G^\circ$ kJ mol^{-1}	S° J K^{-1} mol^{-1}	C_p° J K^{-1} mol^{-1}	$\Delta_{fus}H$ kJ mol^{-1}	$\Delta_{sub}H$ kJ mol^{-1}	p 10^{-30} C m
Sodium (cont.)															
chlorite	$NaClO_2$	90.4	w	2.5	d180		77(3H_2O)	s	−307						
chlorite, hypo-	$NaClO$	74.4	w		d										
	•2$\frac{1}{2}H_2O$	119.5	w		58		$100^{30°C}$								
	•5H_2O	164.5	nil	1.6	18		$64^{23°C}$								
chromate	Na_2CrO_4	162.0	y	2.7	792		85(6H_2O)	s	−1342	−1235	177	142	24		
	•4H_2O	234.0	y		d			s	−2529						
chromate, di-	$Na_2Cr_2O_7$	262.0	r		357	d400		s	−1979						
	•2H_2O	298.0	r	2.3	d100		185	s	−2575						
citrate	$Na_3C_6H_5O_7$	258.1													
	•2H_2O	294.1	w		d150		53								
cobaltinitrite	$Na_3Co(NO_2)_6$	403.9	y				vs	s	−1423						
cyanate	$NaCNO$	65.0	w	1.9	d		hyd	s	−405	−358	97	87			
cyanide	$NaCN$	49.0	w	1.6	564	1496	63(2H_2O)	s	−87	−76	116	70	9	196	
ferricyanide	$Na_3Fe(CN)_6$	280.9													
	•H_2O	298.9	r				$19^{0°C}$								
ferrocyanide	$Na_4Fe(CN)_6$	303.9	y		d435										
	•10H_2O	484.1		1.5	d82		21								
fluoride	NaF 1	42.0	w	2.8	993	1695	4.1	s	−574	−544	51	47	33	282	27.2
formate	$NaCHO_2$	68.0	w	1.9	253	d	100	s	−666	−600	104	83			
hydride	NaH 1	24.0	w	1.4	d425		hyd	s	−56	−33	40	36		187	
hydroxide	$NaOH$	40.0	w	2.1	323	1390		s	−425	−379	64	60	6.4	218	
	•H_2O	58.0	w		d65		114	s	−735	−629	100	90			
iodate	$NaIO_3$	197.9	w	4.3	d			s	−482			92			
	•H_2O	215.9	w				9.4	s	−779	−634	162				
iodate, per-	$NaIO_4$	213.9	w	3.9	d300		14.4	s	−429	−323	163				
iodide	NaI 1	149.9	w	3.7	660	1304		s	−288	−286	99	52	24	208	30.8
	•2H_2O	185.9	w	2.4	d		184	s	−883	−771	196				
molybdate	Na_2MoO_4	205.9	w	3.5	687			s	−1468	−1354	160	142		409	
	•2H_2O	242.0	w	3.3	d100		65	s	−2059	−1830	241				

Substance	*Formula/Structure*	*M* g mol^{-1}	*col.*	*ρ* g cm^{-3}	t_m °C	t_b °C	*sol.* g/100 g	*State*	$\Delta_f H^{\circ}$ kJ mol^{-1}	$\Delta_f G^{\circ}$ kJ mol^{-1}	S° J K^{-1} mol^{-1}	C_p° J K^{-1} mol^{-1}	$\Delta_{fus}H$ kJ mol^{-1}	$\Delta_{sub}H$ kJ mol^{-1}	*p* 10^{-30} C m
Sodium (cont.)															
nitrate	$NaNO_3$	85.0	w	2.3	307	d380	92	s	−468	−367	117	93	16		
nitrite	$NaNO_2$	69.0	w	2.2	271	d320	85	s	−359	−285	104				
nitroprusside	$Na_2[FeNO(CN)_5]$	261.9													
	•$2H_2O$	297.9	r	1.7	d100		$40^{16°C}$								
oxalate	$Na_2C_2O_4$	134.0	w	2.3	d250		3.6	s	−1318			142			
oxide	Na_2O 6	62.0	w	2.3	1132	1275	hyd	s	−414	−375	75	69	48	379	
oxide, per-	Na_2O_2 –	78.0	y	2.8	675	d	hyd	s	−511	−448	95	89			
phosphate	Na_3PO_4	163.9	w	2.5	1340			s	−1917	−1789	174	153			
	•$12H_2O$	380.1	w	1.6	d73		15	s	−5477						
hydrogen	Na_2HPO_4	142.0	w	1.7	d			s	−1748	−1608	150	135			
	•$12H_2O$	358.1	w	1.5	d35		11.8	s	−5298	−4468	634				
dihydrogen	NaH_2PO_4	120.0						s	−1537	−1386	127	117			
	•$2H_2O$	156.0	w	1.9	d60		94	s	−2128						
phosphate															
meta-	$NaPO_3$	102.0	w		d400			s	−1208			92			
	•$2H_2O$	138.0	w		d50		s								
pyro-	$Na_4P_2O_7$	265.9	w	2.5	988			s	−3188	−2969	270	241	58		
	•$10H_2O$	446.0	w	1.8	d94		7.1	s	−6138						
phosphide	Na_3P 9	99.9	r		d		hyd	s	−92						
phosphite															
hydrogen	Na_2HPO_3	126.0						s	−1409						
	•$5H_2O$	216.0	w		53	d200	460	s	−2897						
hypo-	NaH_2PO_2	88.0						s	−839						
	•H_2O	106.0	w		d200		71								
salicylate	$NaC_7H_5O_3$	160.1	w				$100^{20°C}$								
selenate	Na_2SeO_4	188.9	w	3.2				s	−1069						
	•$10H_2O$	369.1	w	1.6	d32		58	s	−4007						
selenide	Na_2Se 6	124.9	w	2.6	>875		hyd	s	−341						
selenite	Na_2SeO_3	172.9	w					s	−959						
	•$5H_2O$	263.0	w				90	s	−2404						

Substance	*Formula/Structure*	M g mol^{-1}	*col.*	ρ g cm^{-3}	t_m °C	t_b °C	*sol.* g/100 g	*State*	$\Delta_f H^\circ$ kJ mol^{-1}	$\Delta_f G^\circ$ kJ mol^{-1}	S° J K^{-1} mol^{-1}	C_p° J K^{-1} mol^{-1}	$\Delta_{fus}H$ kJ mol^{-1}	$\Delta_{sub}H$ kJ mol^{-1}	p 10^{-30} C m
Sodium (cont.)															
silicate	Na_2SiO_3	122.1	w	2.4	1088			s	−1555	−1463	114	112	52		
	•$5H_2O$	212.1	w	1.8	72	d100		s	−3048						
	•$9H_2O$	284.2	w		40	d100	22	s	−4229						
hexafluoro-	Na_2SiF_6	188.1	w	2.7	d		0.75	s	−2910	−2754	207	187			
sulfate	Na_2SO_4	142.0	w	2.7	884	d2227		s	−1387	−1270	150	128	24		
	•$7H_2O$	268.1	w		d24										
	•$10H_2O$	322.2	w	1.5	32	d100	28.0	s	−4327	−3647	592	587			
hydrogen	$NaHSO_4$	120.1	w	2.4	315	d	29	s	−1126	−993	113				
	•H_2O	138.1	w	2.1	d		vs	s	−1422	−1232	155				
peroxydi-	$Na_2S_2O_8$	238.1	w	2.4	d		55								
sulfide	Na_2S 6	78.0	w	1.9	1180			s	−365	−350	84	83	19		
	•$9H_2O$	240.2	w	1.4	(50)	d	20	s	−3074						
sulfite	Na_2SO_3	126.0	w	2.6	d			s	−1101	−1012	146	120			
	•$7H_2O$	252.1	w	1.6	d150		30.0	s	−3162	−2676	444				
hydrogen	$NaHSO_3$	104.1	w	1.5	d										
pyro-	$Na_2S_2O_5$	190.1	w	1.4	d>150		66	s	−1478						
tartrate	$Na_2C_4H_4O_6$	194.0													
	•$2H_2O$	230.1	w	1.8	d150		$45^{17°C}$								
hydrogen	$NaHC_4H_4O_6$	172.1			d238										
	•H_2O	190.1	w		d100		$6.7^{18°C}$								
tellurate															
tetrahydrogen	$Na_2H_4TeO_6$	273.6	w		d		0.8								
telluride	Na_2Te 6	173.6	w	2.9	953		hyd	s	−349	−332	95				
tellurite	Na_2TeO_3	221.6	w				100($5H_2O$)	s	−1003						
thiocyanate	$NaSCN$	81.1	w		287		143($1H_2O$)	s	−171				18		
thionite, di-	$Na_2S_2O_4$	174.1			d		$22^{20°C}$	s	−1232						
thiosulfate	$Na_2S_2O_3$	158.1	w	1.7	d			s	−1123	−1028	155	146			
	•$5H_2O$	248.2	w	1.7	40	d48	76	s	−2608	−2230	372	361	23		
tungstate	Na_2WO_4	293.8	w	4.2	698			s	−1549	−1434	162	140			
	•$2H_2O$	329.9	w	3.2	d100		74	s	−2140						

Substance	*Formula/Structure*	M g mol^{-1}	*col.*	ρ g cm^{-3}	t_m °C	t_b °C	*sol.* g/100 g	*State*	$\Delta_f H^\circ$ kJ mol^{-1}	$\Delta_f G^\circ$ kJ mol^{-1}	S° J K^{-1} mol^{-1}	C_p° J K^{-1} mol^{-1}	$\Delta_{fus}H$ kJ mol^{-1}	$\Delta_{sub}H$ kJ mol^{-1}	p 10^{-30} C m
Sodium (cont.)															
vanadate, ortho-	Na_3VO_4	183.9	w		(850)		22($12H_2O$)	s	−1758	−1638	190	165			
vanadate, meta-	$NaVO_3$	121.9	w		630		21	s	−1146	−1064	114	98			
Strontium	Sr	87.62	s	2.6	769	1384	d	s	0	0	52	26	7.4	164	
	Sr							g	164	131	165	21			
	Sr^{2+}							g	1791						
	Sr^{2+}							aq	−546	−559	−33				
bromide	$SrBr_2$ 9	247.4	w	4.2	643	(2045)		s	−718	−697	135	75	10	308	
	•$6H_2O$	355.5	w	2.4	d88		107	s	−2531	−2174	406	344			
carbonate	$SrCO_3$	147.6	w	3.7	*1497	d1110	0.001	s	−1220	−1140	97	81			
chloride	$SrCl_2$ 6, 9	158.5	w	3.1	874	1250		s	−829	−781	115	76	16	343	
	•$6H_2O$	266.6	w	1.9	d60		56	s	−2624	−2241	391				
chromate	$SrCrO_4$	203.6	y	3.9	d		0.09								
fluoride	SrF_2 6, 9	125.6	w	4.2	1477	2460	0.012	s	−1216	−1165	82	70	30	452	
hydride	SrH_2 9	89.6	w	3.7	d675		hyd	s	−180	−140	50				
hydroxide	$Sr(OH)_2$	121.6	w	3.6	510	d710		s	−969	−881	97	75	21	394	
	•$8H_2O$	265.8	w	1.9	d100		1.0	s	−3352						
iodide	SrI_2 9	341.4	w	4.5	538	d1773		s	−561	−558	159	82	20	286	
	•$6H_2O$	449.5	w	2.7	d90		180	s	−2389			355			
nitrate	$Sr(NO_3)_2$	211.6	w	3.0	570	d1100		s	−978	−780	195	150			
	•$4H_2O$	283.7	w	2.2	d100		80	s	−2155	−1730	369				
nitride	Sr_3N_2 –	290.9			1200		hyd	s	−391	−324	123				
oxide	SrO 1	103.6	w	4.7	2430	(3000)	0.8	s	−592	−562	54	45	70	584	29.7
sulfate	$SrSO_4$	183.7	w	4.0	1605		0.01	s	−1453	−1341	117				
sulfide	SrS 1	119.7	w	3.7	>2000		hyd	s	−472	−468	68	49		581	
Sulfur	S (rhombic)	32.06	y	2.1	113	445	i	s(rhombic, α)	0	0	32	23	1.2		
	S (monoclinic)	32.06	y	2.0	119	445	i	s(monoclinic, β)	0.3	0.1	33	23			
	S							g	277	236	168	24			
	S_2	64.1						g	129	80	228	33			
	S_8	256.5						g	102	50	431	156		102	

Substance	*Formula/Structure*		M g mol^{-1}	*col.*	ρ g cm^{-3}	t_m °C	t_b °C	*sol.* g/100 g	*State*	$\Delta_f H^\circ$ kJ mol^{-1}	$\Delta_f G^\circ$ kJ mol^{-1}	S° J K^{-1} mol^{-1}	C_p° J K^{-1} mol^{-1}	$\Delta_{fus}H$ kJ mol^{-1}	$\Delta_{sub}H$ kJ mol^{-1}	p 10^{-30} C m
Sulfur (cont.)																
acid, sulfamic	NH_2SO_3H		97.1	w	2.1	d200		15	s	−675						
acid, sulfuric	H_2SO_4		98.1	nil	1.8	10	337	∞	l	−814	−690	157	139	10		
sulfate	SO_4^{2-}		96.1						aq	−909	−744	19	−293			
sulfite	SO_3^{2-}		80.1						aq	−635	−487	−29				
peroxydi-	$H_2S_2O_8$		194.1			d65		hyd								
peroxydisulfate	$S_2O_8^{2-}$		192.1						aq	−1345	−1115	244	113	13		
pyro-	$H_2S_2O_7$		178.1	w	1.9	35	d	hyd	s	−1274						
(di-)bromide, di-	S_2Br_2	V	223.9	r	2.6	−46	d>25	hyd	l	−13						
thionyl	$SOBr_2$		207.9	o	2.7	−50	140	hyd	g	−123	−137	333	70			
chloride, di-	SCl_2	V	103.0	r	1.6	−78	d59	hyd	l	−50	−28	184	91		32(vap)	1.2
(di-)chloride, di-	S_2Cl_2	V	135.0	o	1.7	−80	136	hyd	l	−58	−39	224	124		41(vap)	
chloride, tetra-	SCl_4		173.9						l	−57						
sulfuryl	SO_2Cl_2		135.0	nil	1.7	−54	69	hyd	g	−364	−320	312	77		30(vap)	6.0
thionyl	$SOCl_2$		119.0	nil	1.6	−104	76	hyd	g	−212	−198	310	66		33(vap)	4.8
fluoride, hexa-	SF_6	V	146.1	nil	1.9(l)	*−51	s−64	0.004	g	−1209	−1105	292	97	5	14	0
fluoride, tetra-	SF_4	V	108.1	nil		−125	−40	hyd	g	−763	−722	300	78		26(vap)	2.1
sulfuryl	SO_2F_2		102.1	nil	1.7(l)	−136	−55	s	g	−759	−712	284	66			3.7
thionyl	SOF_2		86.1	nil	1.8(l)	−110	−44	hyd	g	−544	−527	279	57			5.4
oxide, di-	SO_2	0	64.1	nil	1.4(l)	−75	−10	9.4	g	−297	−300	248	40	7.4	31	5.5
oxide, tri-	SO_3 (α)	8a	80.1	nil	1.9(l)	62		hyd	l	−441	−374	114		13	58	0
	SO_3 (β)	8a	80.1	nil	2.0	33	45	hyd	g	−396	−371	257	51			
thiosulfate	$S_2O_3^{2-}$		112.1						aq	−652	−522	67				
thiocyanate	SCN^-		58.1						aq	76	92	144	−40			
sulfide	S^{2-}		32.06						aq	33	86	−15				
hydrogen sulfide	SH^-		33.1						aq	−16	12	67				
Tantalum	Ta		180.9	gy	16.6	3000	5458	i	s	0	0	42	25	37	782	
	Ta								g	782	739	185	21			
(V) chloride	$TaCl_5$	0	358.2	y	3.7	216	239	hyd	s	−859	−746	222	148	35	94	
(V) oxide	Ta_2O_5	8a	441.9	w	8.2	1875		i	s	−2046	−1911	143	135	120		

Substance	*Formula/Structure*		M g mol^{-1}	*col.*	ρ g cm^{-3}	t_m °C	t_b °C	*sol.* g/100 g	*State*	$\Delta_f H^{\circ}$ kJ mol^{-1}	$\Delta_f G^{\circ}$ kJ mol^{-1}	S° J K^{-1} mol^{-1}	C_p° J K^{-1} mol^{-1}	$\Delta_{fus}H$ kJ mol^{-1}	$\Delta_{sub}H$ kJ mol^{-1}	p 10^{-30} C m
Tellurium	Te		127.6	s	6.2	450	990	i	s	0	0	50	26	14	87	
	Te								g	197	157	183	21			
	Te_2		255.2						g	168	118	268	37			
acid, telluric	H_6TeO_6		229.6	w	3.1	136		60	s	−1299						
chloride, di-	$TeCl_2$	V	198.5	bk	7.1	208	328	hyd								
chloride, tetra-	$TeCl_4$	9	269.4	w	3.0	224	380	hyd	s	−324	−236	201	138	19	115	
fluoride, hexa-	TeF_6	V	241.6	nil	2.6(l)	*−38	s−39	hyd	g	−1318	−1222	338		8.8	28	0
oxide, di-	TeO_2	7	159.6	w	5.9	733	790	i	s	−323	−270	79	64	29	258	
oxide, tri-	TeO_3	9	175.6	o	5.1	d395		i								
Thallium	Tl		204.4	s	11.8	303	1457	i	s	0	0	64	26	4.3	182	
	Tl								g	182	147	181	21			
	Tl^{+}								g	778						
	Tl^{+}								aq	5	−32	125				
	Tl^{3+}								g	5639						
	Tl^{3+}								aq	197	215	−192				
(I) acetate	$TlC_2H_3O_2$		263.4	w	3.8	131		vs	s	−528						
(I) bromide	TlBr	2	284.3	y	7.6	480	815	0.057	s	−173	−167	120	50	16	136	15.0
(I) chloride	TlCl	2	239.8	w	7.0	430	720	0.39	s	−204	−185	111	51	16	136	15.2
(III) chloride	$TlCl_3$	8iii	310.7	w	4.7	155	d	vs	s	−315	−242	152				
(I) fluoride	TlF	1*	223.4	w	8.2	322	826	$80^{15°C}$	s	−325	−305	96	55	14	142	14.1
(I) hydroxide	TlOH		221.4	y	7.4	d139		$34^{18°C}$	s	−239	−196	88				
(I) iodide	TlI	2, 8v	331.3	y	7.3	440	823	0.008	s	−124	−125	128	52	15	131	15.4
(I) nitrate	$TlNO_3$		266.4	w	5.6	206	430	11.9	s	−244	−152	161	100	10		
(I) oxide	Tl_2O	8ii	424.8	bk	9.5	596	1080	hyd	s	−179	−147	126		30		
(III) oxide	Tl_2O_3	9	456.8	bk	10.2	717	d875	i	s	−390	−302	137	108			
(I) sulfate	Tl_2SO_4		504.8	w	6.8	632	d	5.5	s	−932	−831	230		23		
(I) sulfide	Tl_2S	8ii	440.8	bk	8.5	449	d	0.02	s	−97	−94	151		13		

Substance	*Formula/Structure*		M / g mol^{-1}	*col.*	ρ / g cm^{-3}	t_m / °C	t_b / °C	*sol.* / g/100 g	*State*	$\Delta_f H^\circ$ / kJ mol^{-1}	$\Delta_f G^\circ$ / kJ mol^{-1}	S° / J K^{-1} mol^{-1}	C_p° / J K^{-1} mol^{-1}	$\Delta_{fus} H$ / kJ mol^{-1}	$\Delta_{sub} H$ / kJ mol^{-1}	p / 10^{-30} C m
Thorium	Th		232.0	gy	11.7	1750	4788	i	s	0	0	52	26	14	602	
	Th								g	602	561	190	21			
	Th^{4+}								aq	−769	−705	−423				
chloride	$ThCl_4$	9	373.9	w	4.6	770	922	124(8H_2O)	s	−1187	−1094	190	121			
nitrate	$Th(NO_3)_4$		480.1	w		d500		194(6H_2O)	s	−1441						
	•4H_2O		552.1	w		d			s	−2705						
	•5H_2O		570.1	w		d			s	−3008	−2325	543				
oxide	ThO_2	6	264.0	w	9.9	3390	4400	i	s	−1226	−1169	65	62		728	
sulfate	$Th(SO_4)_2$		424.2	w	4.2	d			s	−2543	−2310	159	172			
	•9H_2O		586.3	w	2.8	d400		1.6								
Tin	Sn (white)		118.7	w	7.3	232	2602	i	s(white)	0	0	51	27	7.1	301	
	Sn (grey)		118.7	gy	5.8			i	s(grey)	−2	0.1	44	26		303	
	Sn								g	301	266	168	21			
(II) bromide	$SnBr_2$	9	278.5	w	5.1	216	620	s	s	−260	−244	150		7.2		
(IV) bromide	$SnBr_4$	0	438.3	w	3.3(l)	30	202	hyd	s	−377	−350	264		12	63	0
(II) chloride	$SnCl_2$	9	189.6	w	3.9	247	635	$270^{15\,°C}$	s	−331	−289	132	79	13		
	•2H_2O		225.6	w	2.7	38	d		s	−921						
(IV) chloride	$SnCl_4$	0	260.5	nil	2.2(l)	−33	114	hyd	l	−511	−440	259	165	9.2	49	0
(IV) hydride	SnH_4	V	122.7	nil		d−150			g	163	188	228	49			0
(II) hydroxide	$Sn(OH)_2$		152.7	w		d		0.0002	s	−561	−492	155				
(IV) hydroxide	$Sn(OH)_4$		186.7						s	−1110						
(II) iodide	SnI_2	9	372.5	o	5.3	320	717	1.1	s	−143				19		
(IV) iodide	SnI_4	0	626.3	o	4.5	145	365	hyd	g			446	105	19		0
(II) oxide	SnO	8v	134.7	bk	6.4	d1080		i	s	−281	−252	57	44		301	14.4
(IV) oxide	SnO_2	7	150.7	w	7.0		s1800	i	s	−578	−516	49	53			
(II) sulfate	$SnSO_4$		214.8	w	4.2	d378										
(II) sulfide	SnS	1*	150.8	gy	5.2	882	1230	i	s	−100	−98	77	49	32	219	10.6
(IV) sulfide	SnS_2	8ii	182.8	y	4.5	d600		i	s	−154	−145	87	70			

Substance	*Formula/Structure*		M / g mol^{-1}	*col.*	ρ / g cm^{-3}	t_m / °C	t_b / °C	*sol.* / g/100 g	*State*	$\Delta_f H^⊖$ / kJ mol^{-1}	$\Delta_f G^⊖$ / kJ mol^{-1}	$S^⊖$ / J K^{-1} mol^{-1}	$C_p^⊖$ / J K^{-1} mol^{-1}	$\Delta_{fus}H$ / kJ mol^{-1}	$\Delta_{sub}H$ / kJ mol^{-1}	p / 10^{-30} C m
Tin (cont.)																
tetraethyl	$Sn(C_2H_5)_4$		235.0	nil	1.2	−112	181	i	l	−96					50(vap)	
tetramethyl	$Sn(CH_3)_4$		178.8	nil	1.3	−55	78	i	l	−52					34(vap)	
Titanium	Ti		47.87	s	4.5	1660	3287	i	s	0	0	31	25	15	473	
	Ti								g	473	428	180	24			
(IV) bromide	$TiBr_4$	0	367.5	y	3.3	39	230	hyd	s	−617	−590	244	131	13	68	0
(IV) carbide	TiC	1	59.9	gy	4.9	3140	4820	i	s	−185	−181	24	34	71		
(II) chloride	$TiCl_2$	8ii	118.8	bk	3.1	1025	d	hyd	s	−514	−464	87	70		212	
(III) chloride	$TiCl_3$	8iv	154.2	v	2.6	d440		s	s	−721	−654	140	97	21	179	
(IV) chloride	$TiCl_4$	V	189.7	w	1.7(l)	−24	136	s	l	−804	−737	252	145	9.4	50	0
(II) hydride	TiH_2		49.9	gy	3.8	d450		i	s	−120	−80	30	30			
(III) nitride	TiN	1	61.9	y	5.2	2930		i	s	−338	−310	30	37	67		
(II) oxide	TiO	1	63.9	y	5.0	1770	3227		s	−543	−513	35	40	42		
(IV) oxide	TiO_2	7	79.9	w	4.2	(1830)	(2500)	i	s(rutile)	−944	−890	51	55	65	694	
Tungsten	W		183.8	gy	19.3	3410	5660	i	s	0	0	33	24	35	849	
	W								g	849	807	174	21			
acid, tungstic	H_2WO_4		249.9	y	5.5	d100		i	s	−1132						
carbide	WC	9	195.9	gy	15.7	2870	6000	i	s	−41	−42	42	36			
(di-) carbide	W_2C	9	379.7	gy	17.2	2860	6000	i	s	−26						
carbonyl	$W(CO)_6$		351.9	w	2.6	d170		i	s	−954						
(IV) chloride	WCl_4	8a	325.7	gy	4.6	d		hyd	s	−443	−359	198	130	6.4	107	
(VI) chloride	WCl_6	0	396.6	b	3.5	275	347	hyd	s	−594	−455	238	175	6.7	100	0
(IV) oxide	WO_2	7	215.9	bn	12.1	1550	d1724	i	s	−590	−534	51	56		636	
(VI) oxide	WO_3	9	231.9	y	7.2	1473	1837	i	s	−843	−764	76	74	73	546	
Uranium	U		238.0	s	19.1	1135	3818	i	s	0	0	50	28	15	533	
	U								g	533	488	200	24			
acetate, uranyl	$UO_2(C_2H_3O_2)_2$		388.1	y		d275										
	$\bullet 2H_2O$		424.2	y	2.9	d110		$8.3^{17°C}$	s	−2559						

Substance	*Formula/Structure*		M / g mol^{-1}	*col.*	ρ / g cm^{-3}	t_m / °C	t_b / °C	*sol.* / g/100 g	*State*	$\Delta_f H^\circ$ / kJ mol^{-1}	$\Delta_f G^\circ$ / kJ mol^{-1}	S° / J K^{-1} mol^{-1}	C_p° / J K^{-1} mol^{-1}	$\Delta_{fus}H$ / kJ mol^{-1}	$\Delta_{sub}H$ / kJ mol^{-1}	p / 10^{-30} C m
Uranium (cont.)																
chloride, uranyl	UO_2Cl_2		340.9	y	5.4	d577		320$^{18°C}$	s	−1244	−1146	151	108			
fluoride, hexa-	UF_6	V	352.0	w	5.1	*65	s57	hyd	s	−2197	−2068	228	167	19	50	0
formate, uranyl	$UO_2(CHO_2)_2$		360.1	y					s	−1850						
	•H_2O		378.1	y	3.7	d110		8.2	s	−2157						
nitrate, uranyl	$UO_2(NO_3)_2$		394.0						s	−1349	−1105	243				
	•$6H_2O$		502.1	y	2.8	60	d100	127	s	−3168	−2585	506	467			
(IV) oxide	UO_2	6	270.0	bn	11.0	2865		i	s	−1085	−1032	77	64		619	
(VI) oxide	UO_3	9	286.0	y	7.3	d1300		i	s	−1224	−1146	96	82			
(IV/VI) oxide (pitchblende)	U_3O_8	9	842.1	g	8.3	d1300		i	s	−3575	−3370	283	238			
sulfate, uranyl	UO_2SO_4		366.1						s	−1845	−1684	155	145			
	•$3H_2O$		420.1	y	3.3	d100		158	s	−2754	−2418	268	283			
Vanadium	V		50.94	s	6.1	1910	3407	i	s	0	0	29	25	18	514	
	V								g	514	469	182	26			
(II) chloride	VCl_2	8ii	121.9	g	3.2	1350		hyd	s	−452	−406	97	72		196	
(III) chloride	VCl_3	8iv	157.3	v	3.0	d		hyd	s	−581	−511	131	93			
(IV) chloride	VCl_4	V	192.8	r	1.8	−28	149	hyd	l	−569	−504	255		2.3	46	0
(II) oxide	VO	9	66.9	gy	5.6	1790		i	s	−432	−404	39	45	63	536	
(III) oxide	V_2O_3	9	149.9	bk	4.9	1970		i	s	−1219	−1139	98	103			
(IV) oxide	VO_2	7*	82.9	b	4.3	1967		i	s	−713	−659	51	59	57		
(V) oxide	V_2O_5	8v	181.9	o	3.4	690	d1750	0.07	s	−1551	−1420	131	128	65		
(IV) sulfate, vanadyl	$VOSO_4$		163.0	b				s	s	−1309	−1170	109				
	•$5H_2O$		253.1													
Xenon	Xe		131.3	nil	3.5(l)	−112	−108	0.058$^{20°C}$	g	0	0	170	21	2.3		0
fluoride, di-	XeF_2	0	169.3	w	4.3	*129	s114	2.5$^{0°C}$	g	−130	−96	260			52	0
fluoride, tetra-	XeF_4	0	207.3	w	4.0	*117	s116	hyd	g	−215	−138	316	90		63	0
fluoride, hexa-	XeF_6	9	245.3	w	3.4	50	76	hyd	g	−294					38	0

Substance	*Formula/Structure*		M g mol^{-1}	*col.*	ρ g cm^{-3}	t_m °C	t_b °C	*sol.* g/100 g	*State*	$\Delta_f H^\circ$ kJ mol^{-1}	$\Delta_f G^\circ$ kJ mol^{-1}	S° J K^{-1} mol^{-1}	C_p° J K^{-1} mol^{-1}	$\Delta_{fus}H$ kJ mol^{-1}	$\Delta_{sub}H$ kJ mol^{-1}	p 10^{-30} C m
Xenon (cont.)																
oxide, tri-	XeO_3	V	179.3	w	4.6	d40		hyd	g	502	561	287	62		100	
Ytterbium	Yb		173.1	s	7.0	824	1194	hyd	s	0	0	60	27	7.7		
	Yb								g	152	118	173	21			
	Yb^{3+}								g	4367						
	Yb^{3+}								aq	−674	−644	−238	25			
(III) fluoride	YbF_3	9	230.1	w	8.2	1157	2230	i								
(III) oxide	Yb_2O_3		394.2	w	9.2	2355	4070		s	−1815	−1727	133	115			
Yttrium	Y		88.91	s	4.5	*1522	3338	hyd	s	0	0	44	27	11.0		
	Y								g	421	381	179	26			
	Y^{3+}								g	4200						
	Y^{3+}								aq	−723	−694	−251				
fluoride	YF_3	9	145.9	w	4.0	1155	2230	i	s	−1719	−1645	100	70(g)	28		
nitrate	$Y(NO_3)_3$		274.9													
	$\cdot 6H_2O$		383.0	w	2.7	d100		$134^{22°C}$								
oxide	Y_2O_3		225.8	w	5.0	2420	4300	i	s	−1905	−1817	99	103	105		
Zinc	Zn		65.38	gy	7.1	420	907	i	s	0	0	42	25	7.4	130	
	Zn								g	130	94	161	21			
	Zn^{2+}								g	2783						
	Zn^{2+}								aq	−153	−147	−110	46			
acetate	$Zn(C_2H_3O_2)_2$		183.5	w	1.8	d200			s	−1079						
	$\cdot 2H_2O$		219.5	w	1.7	d100		35	s	−1672						
acetylacetonate	$Zn(C_5H_7O_2)_2$		263.6	y		124		$0.4^{20°C}$								
bromide	$ZnBr_2$	8i	225.2	w	4.2	394	650		s	−329	−312	138	66	16	131	
	$\cdot 2H_2O$		261.2			37	d100	470	s	−937	−799	199				
carbonate	$ZnCO_3$		125.4	w	4.4	d140		0.021	s	−813	−732	82	80			
hydroxide	$ZnCO_3$		324.1													
	$\cdot 2Zn(OH)_2 \cdot H_2O$[h]		342.2	w	3.5	d140		i								
chloride	$ZnCl_2$	8v, 9, 9	136.3	w	2.9	283	732	432	s	−415	−369	111	71	10	149	

[h] ($ZnCO_3 \cdot 2Zn(OH)_2 \cdot H_2O$) – approximate formula of the precipitated carbonate.

Substance	*Formula/Structure*		M g mol^{-1}	*col.*	ρ g cm^{-3}	t_m °C	t_b °C	*sol.* g/100 g	*State*	$\Delta_f H^\circ$ kJ mol^{-1}	$\Delta_f G^\circ$ kJ mol^{-1}	S° J K^{-1} mol^{-1}	C_p° J K^{-1} mol^{-1}	$\Delta_{fus}H$ kJ mol^{-1}	$\Delta_{sub}H$ kJ mol^{-1}	p 10^{-30} C m
Zinc (cont.)																
chromate	$ZnCrO_4$		181.4	y	3.4	316		i								
cyanide	$Zn(CN)_2$		117.4	w	1.9	d800		0.006$^{18°C}$	s	96						
fluoride	ZnF_2	7	103.4	w	5.0	872	1500		s	−764	−713	74	66	42		
	$\bullet 4H_2O$		175.4	w	2.3	d100		1.5								
hydroxide	$Zn(OH)_2$		99.4	w	3.1	d125		0.001	s	−642	−554	81				
iodide	ZnI_2	8ii	319.2	w	4.7	446	d624	435	s	−208	−209	161			120	
nitrate	$Zn(NO_3)_2$		189.4	w		d			s	−484						
	$\bullet 6H_2O$		297.5	w	2.1	36	d105	127	s	−2307	−1773	457	323			
nitride	Zn_3N_2	9	224.1	gy	6.2	d700		hyd	s	−23	30	140	109			
oxide	ZnO	4	81.4	w	5.6	1975		0.001	s	−350	−320	44	40			
sulfate	$ZnSO_4$		161.4	w	3.5	d680			s	−983	−872	110	99			
	$\bullet 6H_2O$		269.5	w	2.1	d70			s	−2777	−2324	364	358			
	$\bullet 7H_2O$		287.5	w	2.0	d100		58	s	−3078	−2563	389	383			
sulfide	ZnS (wurtzite)	4	97.4	w	4.0	*1722	s1185	i	s(wurtzite)	−193	−191	68	46		289	
	ZnS (zinc blende)	3	97.4	w	4.1	d1020		i	s(zinc blende)	−206	−201	58	46			
Zirconium	Zr		91.2	s	6.5	1852	4377	i	s	0	0	39	26	19	609	
	Zr								g	609	567	181	27			
chloride	$ZrCl_4$	9	233.0	w	2.8	*437	s331	s	s	−981	−890	182	120	38	110	0
chloride,	$ZrOCl_2$		178.1	w		d250		s								
zirconyl	$\bullet 8H_2O$		322.3		1.9	d210		vs	s	−3472						
oxide	ZrO_2	9	123.2	w	5.9	2678	4300	i	s	−1101	−1043	50	56	87	787	

17 SOME CRYSTAL FORMS

Numbers before the name refer to the structural types listed in the introduction to table 16: 'Properties of elements and inorganic compounds'.

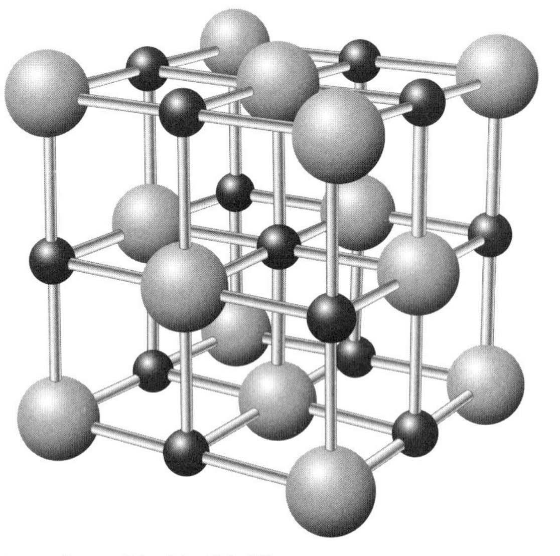

1. sodium chloride (NaCl)
6:6 coordination
(● Na, ○ Cl)

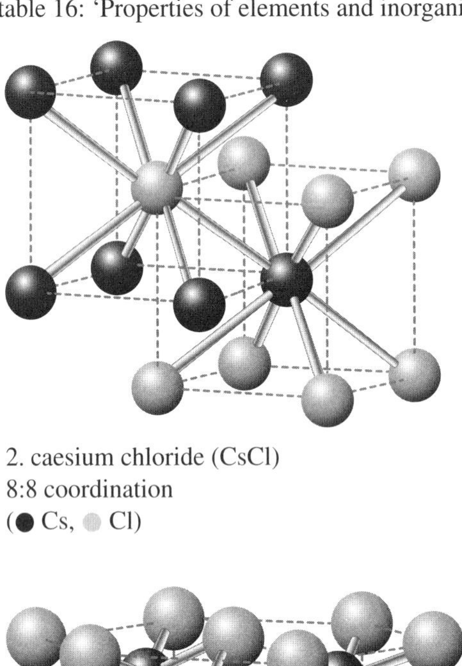

2. caesium chloride (CsCl)
8:8 coordination
(● Cs, ○ Cl)

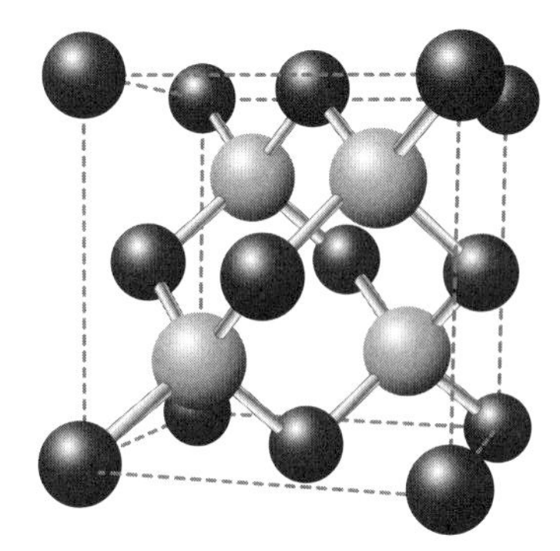

3. zinc blende (ZnS)
4:4 coordination
(● Zn, ○ S)

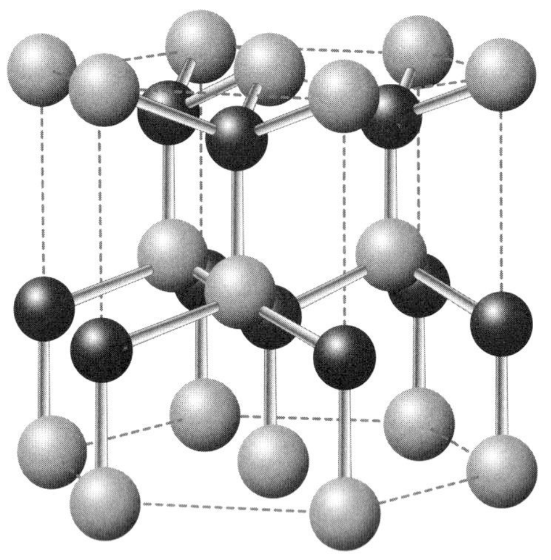

4. wurtzite (ZnS)
4:4 coordination
(● Zn, ○ S)

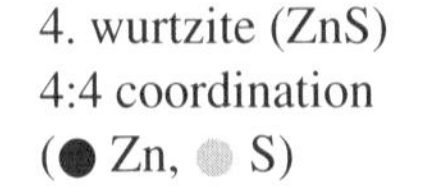

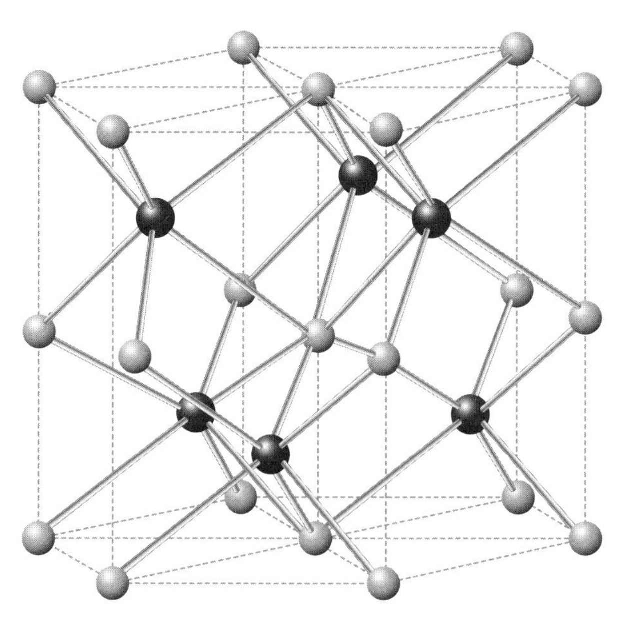

5. nickel arsenide (NiAs)
6:6 coordination
(● As, ● Ni)

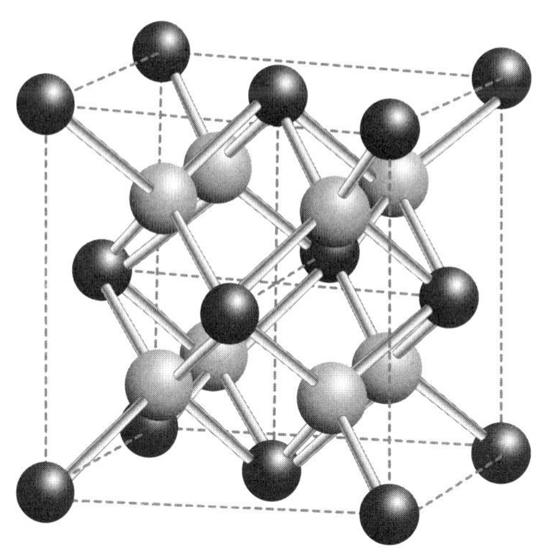

6. fluorite (CaF_2)
8:4 coordination
(● Ca, ● F)

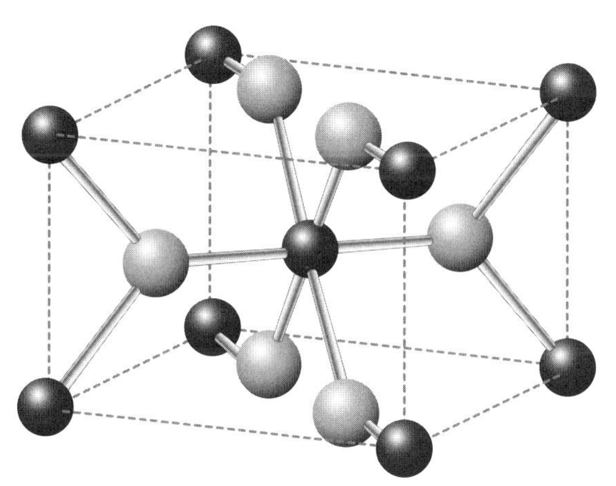

7. rutile (TiO_2)
6:3 coordination
(● Ti, ● O)

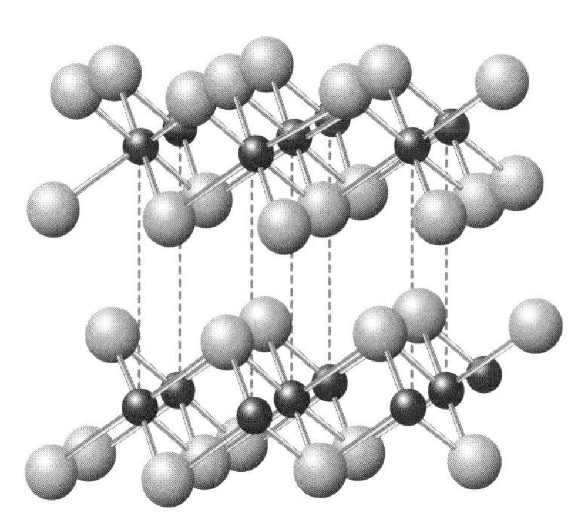

8i. cadmium chloride ($CdCl_2$)
layer structure
(● Cd, ● Cl)

8ii. cadmium iodide (CdI_2)
layer structure
(● Cd, ● I)

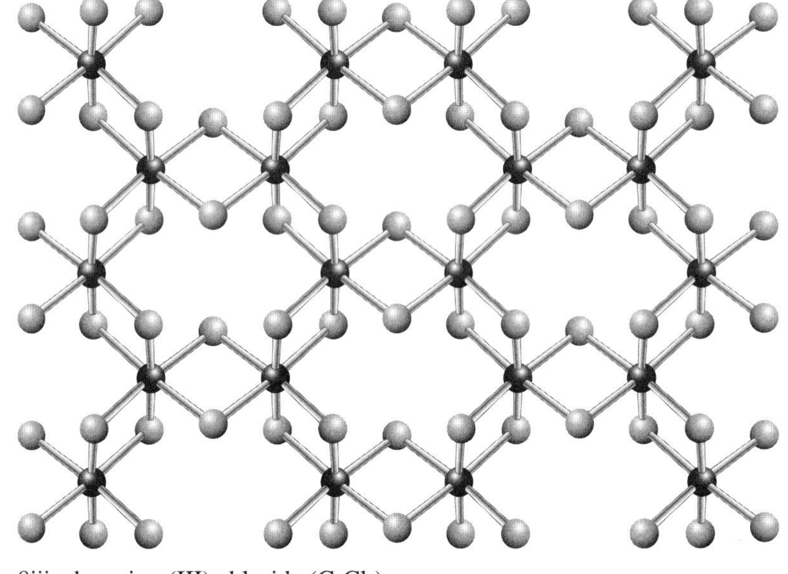

8iii. chromium(III) chloride ($CrCl_3$)
layer structure
(● Cr, ● Cl)

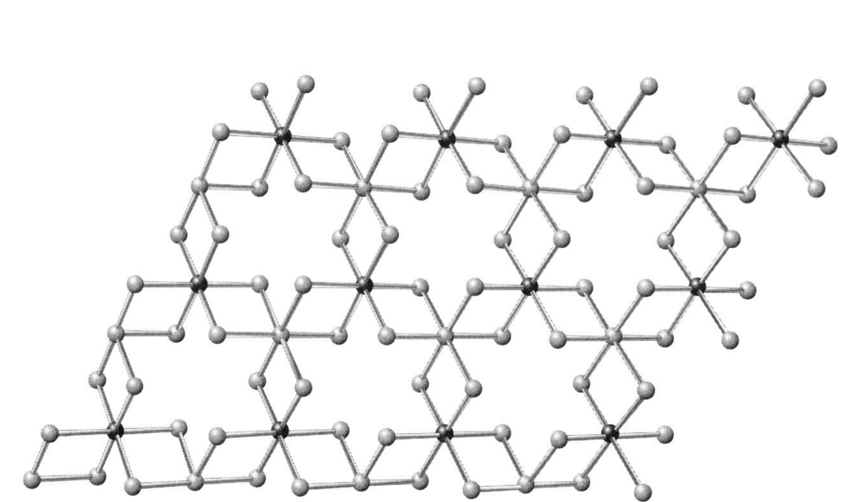

8iv. bismuth iodide (BiI_3)
layer structure
(● Bi, ● I)

18 SHAPES OF SOME MOLECULES AND IONS

Substances are mostly listed in alphabetical order of the symbol of the central atom. A is the central atom bonded to B, and E is a non-bonding electron pair. Bond angles (°) normally apply to a molecule in the gas phase, or to an ion in a crystal lattice. Symmetry labels refer to the idealised geometries.

Shape	Examples		
Linear	$Ag(CN)_2^-$	HgX_2	N_3^-
(AB_2 or AB_2E_3)	CO_2	I_3^-	XeF_2
	CS_2	KrF_2	
Trigonal planar	BX_3 (X = Br,Cl,F,I)	GaI_3	SO_3
(120°)	CO_3^{2-}	NO_3^-	
(AB_3)			
Bent	$(CH_3)O(CH_3)$ (111°)	H_2Se (91°)	SCl_2 (100°)
(AB_2E or AB_2E_2)	Cl_2O (111°)	H_2Te (89°)	SO_2 (120°)
	ClO_2 (117°)	NO_2 (134°)	SnX_2
	ClO_2^- (108°)	NO_2^- (115°)	(X = Br,Cl,I)
	H_2O (105°)	O_3 (117°)	
	H_2S (92°)	OF_2 (103°)	
Tetrahedral	$AlCl_4^-$	GeH_4	SeO_4^{2-}
(109.5°)	AsO_4^{3-}	MnO_4^-	SiX_4
(AB_4)	BF_4^-	NH_4^+	(X = Br,Cl,F,H,I)
	BH_4^-	$Ni(CO)_4$	SnX_4
	BeF_4^{2-}	PCl_4^+	(X = Br,Cl,H,I)
	CX_4	PH_4^+	$TiBr_4$
	(X = Br,Cl,F,H,I)	PO_4^{3-}	$TiCl_4$
	ClO_4^-	SO_4^{2-}	VCl_4
	CrO_4^{2-}		$Zn(CN)_4^{2-}$
	$Cu(CN)_4^{3-}$		$ZrCl_4$

Shape	Examples		
Trigonal pyramidal	$AsBr_3$ (100°)	$N(CH_3)_3$ (111°)	PH_3 (94°)
(AB_3E)	$AsCl_3$ (99°)	NCl_3 (107°)	SO_3^{2-}
	AsF_3 (96°)	NF_3 (102°)	$SbBr_3$ (98°)
	AsH_3 (92°)	NH_3 (107°)	$SbCl_3$ (97°)
	AsI_3 (100°)	PBr_3 (101°)	SbF_3 (87°)
	$BiBr_3$ (100°)	$P(CH_3)_3$ (99°)	SbH_3 (91°)
	$BiCl_3$ (100°)	PCl_3 (100°)	SbI_3 (99°)
	ClO_3^- (107°)	PF_3 (98°)	XeO_3
Trigonal bipyramidal	AsF_5	PF_5	
(90° and 120°) (AB_5)	PCl_5	$SbCl_5$	
Seesaw (AB_4E)	SF_4		
T-shaped (AB_3E_2)	BrF_3 (86°)	ClF_3 (87°)	
Octahedral	MoF_6	SeF_6	WCl_6
(90°)	PCl_6^-	TeF_6	XeF_6
(AB_6)	SF_6	UF_6	
	Six-coordinated complexes of many metals		
Square pyramidal	BrF_5	IF_5	
(AB_5E)	ClF_5	$XeOF_4$	
Square planar	$AuCl_4^-$	$Ni(CN)_4^{2-}$	$Pt(NH_3)_4^{2+}$
(90°) (AB_4E_2)	ICl_4^-	$PdCl_4^{2-}$	XeF_4

SHAPES OF MOLECULES

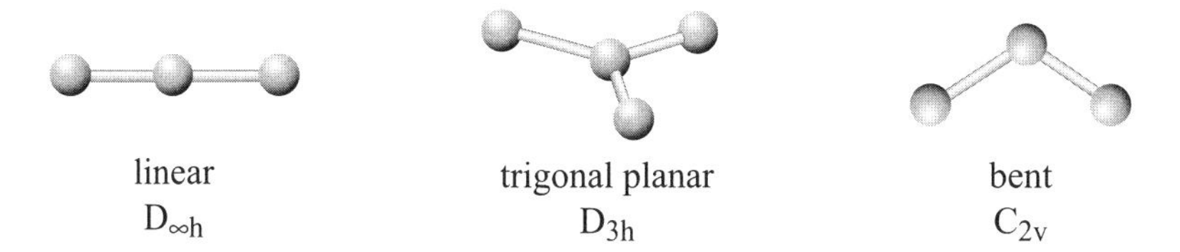

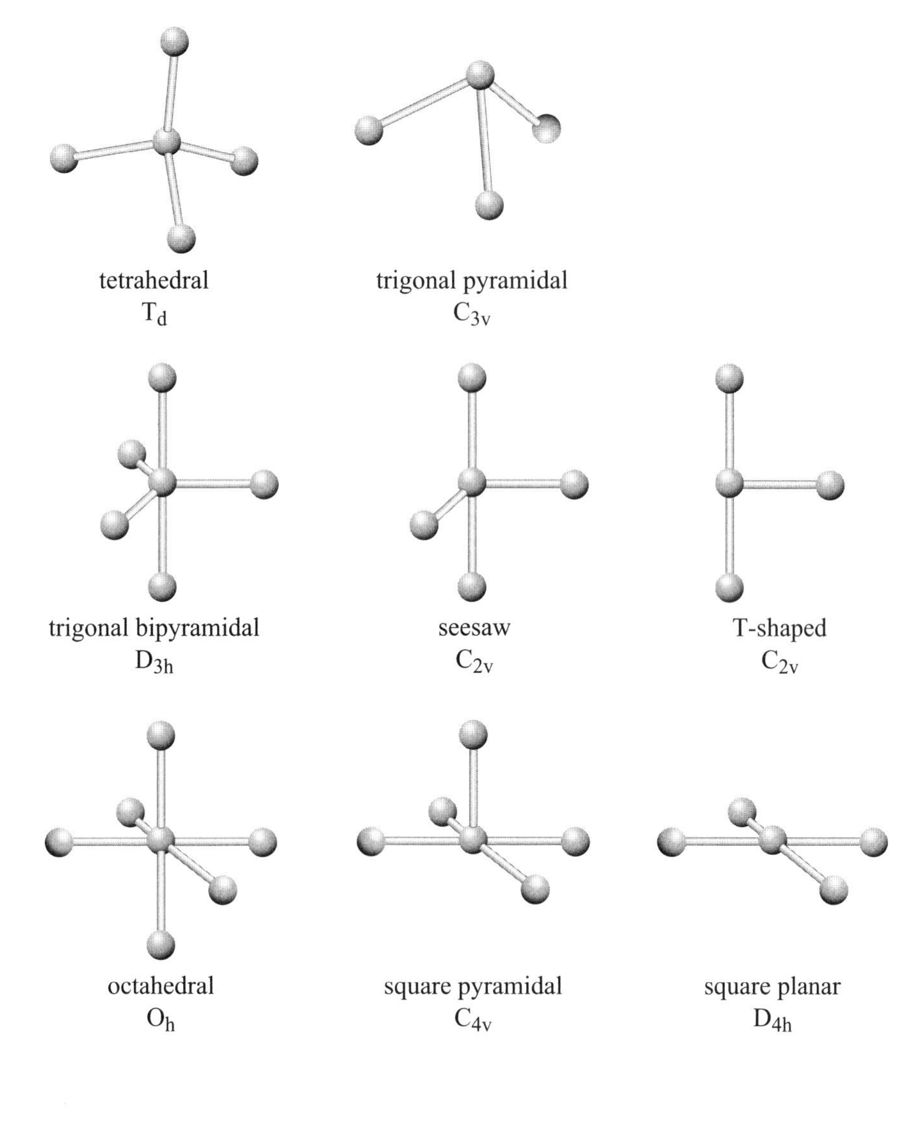

19 BOND LENGTHS

Bond lengths are given in pm. 1 pm = 10^{-12} m = 0.01 Å. In many cases the bond lengths are average values for molecules containing the bonds.

Bond length/pm

Single bonds

	Br	C	Cl	F	H	I	N	O	P	S	Si
Br	228	194	214	176	141	247	214		220	227	216
C	194	154	177	138	108	214	147	143	184	182	185
Cl	214	177	199	163	128	232	197	170	203	199	202
F	176	138	163	142	92	257	136	142	154	158	156
H	141	108	128	92	74	160	101	97	142	134	148
I	247	214	232	257	160	267			247		243
N	214	147	197	136	101		146	136		175	174
O		143	170	142	97		136	148	154	161	163
P	220	184	203	154	142	247		154	221	210	
S	227	182	199	158	134		175	161	210	205	215
Si	216	185	202	156	148	243	174	163		215	232

Multiple bonds

C=C	134	C≡N	116	N≡N	110
C≡C	120	C=O	122	N=O	114
C⩦C (benzene)	140	C=S	156	O=O	121
C=N	130	N=N	125	S=S	189

20 AVERAGE BOND ENTHALPIES

The bond enthalpy is the enthalpy change for the breaking of a mole of bonds. For diatomic molecules this is the enthalpy change for the dissociation

$$XY(g) \longrightarrow X(g) + Y(g)$$

For polyatomic molecules, XY_n, which contain a single type of bond, the average bond enthalpy is $\Delta H/n$ for

$$XY_n(g) \longrightarrow X(g) + nY(g)$$

For polyatomic molecules containing more than one type of bond, average bond enthalpies are estimated from the enthalpies of dissociation of two or more compounds containing some bonds in common; for example, the average C—C and C—H bond enthalpies are based on the enthalpies of dissociation of a number of alkanes. All data refer to 298 K.

ΔH/kJ mol^{-1}

Single bonds

	Br	C	Cl	F	H	I	N	O	P	S	Si
Br	193	285	219	249	366	178		201	264	218	330
C	285	346	324	492	414	228	286	358	264	289	307
Cl	219	324	242	255	431	211	192	206	322	271	400
F	249	492	255	159	567	280	278	191	490	327	597
H	366	414	431	567	436	298	391	463	322	364	323
I	178	228	211	280	298	151		201	184		234
N		286	192	278	391		158	214			
O	201	358	206	191	463	201	214	144	363		466
P	264	264	322	490	322	184		363	198		
S	218	289	271	327	364					266	293
Si	330	307	400	597	323	234		466		293	226

Multiple bonds

C=C	614		C=N	615		N=N	470		O=O	498
C≡C	839		C≡N	890		N≡N	945		S=S	429
			C=O	804		N=O	587			
			C=S	536						

21 LATTICE ENTHALPIES OF IONIC CRYSTALS

The lattice energy of an ionic crystal is the change in internal energy at 0 K, $\Delta U_{0\,K}$, which accompanies the dissociation of one mole of the crystal into its constituent gaseous ions. The process may be represented as, for example for the alkali halides:

$$MX(s) \longrightarrow M^+(g) + X^-(g)$$

Below, we tabulate the enthalpy change at 298 K (25 °C) for the above process. It is related to $\Delta U_{0\,K}$ by the equation

$$\Delta H_{298\,K} = \Delta U_{0\,K} + \int_{0\,K}^{298\,K} [C_p(M^+) + C_p(X^-) - C_p(MX)]\,dT$$

The values for the halides have been calculated from thermochemical data at 298 K using a Born-Haber cycle. The values for the oxides and sulfides are based on the calculated values quoted by Jenkins (1978).

$\Delta H_{298\,K}$/kJ mol^{-1}

	F	Cl	Br	I	O	S
Li	1047	862	818	759	2806	2471
Na	928	788	751	700	2488	2199
K	826	718	689	645	2245	1986
Rb	793	693	666	627	2170	1936
Cs	756	668	645	608		1899
Be	3509	3017	2909	2792	4298	3846
Mg	2961	2523	2434	2318	3800	3323
Ca	2634	2255	2170	2065	3419	3043
Sr	2496	2153	2070	1955	3222	2879
Ba	2357	2053	1980	1869	3034	2716

22 STABILITY CONSTANTS OF COMPLEX IONS

The equilibrium

$$[ML_{n-1}]^{x+}(aq) + L(aq) \rightleftharpoons [ML_n]^{x+}(aq)$$

where M indicates a metal ion and L a ligand, has a stepwise stability constant (K_n)

$$K_n = \frac{[ML_n^{x+}(aq)]}{[ML_{n-1}^{x+}(aq)][L(aq)]^n}$$

For the formation of a metal complex $[ML_n]^{x+}$ from a metal ion M^{x+} and the ligands, nL, in aqueous solution according to the equation

$$M^{x+}(aq) + nL(aq) \rightleftharpoons [ML_n]^{x+}(aq)$$

an equilibrium constant called the cumulative formation constant (β_n) can be defined as

$$\beta_n = \frac{[ML_n^{x+}(aq)]}{[M^{x+}(aq)][L(aq)]^n}$$

Note:

$K_1 \times K_2 \times \ldots \times K_n = \beta_n$, and $\log_{10}K_1 + \log_{10}K_2 + \ldots + \log_{10}K_n = \log_{10}\beta_n$ where K_1, K_2 etc. are the stepwise formation constants and β_n the cumulative formation constant.

Equilibrium	$\log_{10}\beta_n$
$Ag^+ + 2CN^- \rightleftharpoons [Ag(CN)_2]^-$	20.48
$AgCl(s) + Cl^- \rightleftharpoons [AgCl_2]^-$	−4.50
$AgI(s) + I^- \rightleftharpoons [AgI_2]^-$	−4.4[a]
$AgI(s) + 2I^- \rightleftharpoons [AgI_3]^{2-}$	−3.0[a]
$Ag^+ + 2NH_3 \rightleftharpoons [Ag(NH_3)_2]^+$	7.22
$Ag^+ + 2S_2O_3^{2-} \rightleftharpoons [Ag(S_2O_3)_2]^{3-}$	13.5
$Al^{3+} + 6F^- \rightleftharpoons [AlF_6]^{3-}$	20
$Ba^{2+} + EDTA^{4-} \rightleftharpoons [Ba(EDTA)]^{2-}$	7.86
$Ca^{2+} + EBT \rightleftharpoons [Ca(EBT)]^{2+}$	5.4
$Ca^{2+} + EDTA^{4-} \rightleftharpoons [Ca(EDTA)]^{2-}$	10.65
$Cd^{2+} + 4CN^- \rightleftharpoons [Cd(CN)_4]^{2-}$	17.92
$Cd^{2+} + 4I^- \rightleftharpoons [CdI_4]^{2-}$	6.0
$Cd^{2+} + 4NH_3 \rightleftharpoons [Cd(NH_3)_4]^{2+}$	6.74
$Co^{2+} + 6NH_3 \rightleftharpoons [Co(NH_3)_6]^{2+}$	4.4[b]
$Co^{3+} + 6NH_3 \rightleftharpoons [Co(NH_3)_6]^{3+}$	34.4
$Co^{2+} + EDTA^{4-} \rightleftharpoons [Co(EDTA)]^{2-}$	16.45
$Co^{3+} + EDTA^{4-} \rightleftharpoons [Co(EDTA)]^-$	41.4
$Co^{2+} + 3en \rightleftharpoons [Co(en)_3]^{2+}$	14.0
$Co^{3+} + 3en \rightleftharpoons [Co(en)_3]^{3+}$	48.7
$Cu^+ + 4CN^- \rightleftharpoons [Cu(CN)_4]^{3-}$	28
$Cu^+ + 2NH_3 \rightleftharpoons [Cu(NH_3)_2]^+$	10.6
$Cu^{2+} + 4NH_3 \rightleftharpoons [Cu(NH_3)_4]^{2+}$	11.75
$Cu^{2+} + EDTA^{4-} \rightleftharpoons [Cu(EDTA)]^{2-}$	18.78
$Cu^{2+} + 2en \rightleftharpoons [Cu(en)_2]^{2+}$	19.6
$Cu^{2+} + 4C_5H_5N \rightleftharpoons [Cu(C_5H_5N)_4]^{2+}$	6.0
$Fe^{2+} + 6CN^- \rightleftharpoons [Fe(CN)_6]^{4-}$	35.4
$Fe^{3+} + 6CN^- \rightleftharpoons [Fe(CN)_6]^{3-}$	43.6
$Fe^{3+} + SCN^- \rightleftharpoons [Fe(SCN)]^{2+}$	3.02
$Fe^{2+} + EDTA^{4-} \rightleftharpoons [Fe(EDTA)]^{2-}$	14.30
$Fe^{3+} + EDTA^{4-} \rightleftharpoons [Fe(EDTA)]^-$	25.1
$Fe^{2+} + 3phen \rightleftharpoons [Fe(phen)_3]^{2+}$	20.7
$Fe^{3+} + 3phen \rightleftharpoons [Fe(phen)_3]^{3+}$	13.8
$Hg^{2+} + 4CN^- \rightleftharpoons [Hg(CN)_4]^{2-}$	39.0
$Hg^{2+} + 4Cl^- \rightleftharpoons [HgCl_4]^{2-}$	15.6
$Hg^{2+} + 4I^- \rightleftharpoons [HgI_4]^{2-}$	29.8
$Mg^{2+} + EBT \rightleftharpoons [Mg(EBT)]^{2+}$	7.0
$Mg^{2+} + EDTA^{4-} \rightleftharpoons [Mg(EDTA)]^{2-}$	8.85

Equilibrium	$\log_{10}\beta_n$
$Mn^{3+} + 3C_2O_4^{2-} \rightleftharpoons [Mn(C_2O_4)_3]^{3-}$	18.4
$Ni^{2+} + 4CN^- \rightleftharpoons [Ni(CN)_4]^{2-}$	30.22
$Ni^{2+} + 6NH_3 \rightleftharpoons [Ni(NH_3)_6]^{2+}$	8.31
$Pb^{2+} + EDTA^{4-} \rightleftharpoons [Pb(EDTA)]^{2-}$	18.0
$Zn^{2+} + 4CN^- \rightleftharpoons [Zn(CN)_4]^{2-}$	19.62
$Zn^{2+} + 4NH_3 \rightleftharpoons [Zn(NH_3)_4]^{2+}$	8.89
$Zn^{2+} + EBT \rightleftharpoons [Zn(EBT)]^{2+}$	12.3
$Zn^{2+} + EDTA^{4-} \rightleftharpoons [Zn(EDTA)]^{2-}$	16.5
$Zn^{2+} + 3en \rightleftharpoons [Zn(en)_3]^{2+}$	13.9

Unless otherwise stated, all constituents are in aqueous solution at 298 K and concentrations are expressed in mol L^{-1}.
[a] 291 K
[b] 293 K
EBT = eriochrome black T
$EDTA^{4-}$ = ethylenediaminetetraacetato ion
en = ethylenediamine
phen = 1,10-phenanthroline
C_5H_5N = pyridine

23 SOLUBILITY PRODUCTS

For a solid M_aL_b in equilibrium with its ions in aqueous solution,

$$M_aL_b(s) \rightleftharpoons aM^{b+} + bL^{a-}$$

the solubility product is the product of the concentrations

$$K_{sp} = [M^{b+}]^a[L^{a-}]^b$$

each expressed in mol L^{-1}. In some cases the solid that is in equilibrium with the solution is a hydrate. All data refer to 298 K except where noted.

Solid	K_{sp}
AgBr	5.0×10^{-13}
Ag_2CO_3	8.1×10^{-12}
$AgC_2H_3O_2$	2×10^{-3}
AgCl	1.8×10^{-10}
Ag_2CrO_4	2.6×10^{-12}
AgI	8.3×10^{-17}
Ag_3PO_4	2.6×10^{-18}
Ag_2S	8×10^{-51}
AgSCN	1.1×10^{-12}
Ag_2SO_4	1.5×10^{-5}
$Al(OH)_3$	3×10^{-34}
$BaCO_3$	2.0×10^{-9}
$BaCrO_4$	2.1×10^{-10}
$Ba(OH)_2$	3×10^{-4}
$BaSO_4$	1.1×10^{-10}
$Be(OH)_2$	1×10^{-21}
$CaCO_3$	3.3×10^{-9}
CaC_2O_4	4×10^{-9} (335 K)
CaF_2	3.2×10^{-11}
$Ca(OH)_2$	6.4×10^{-6}
$Ca_3(PO_4)_2$	1.2×10^{-29}
$CaSO_4$	2.4×10^{-5}
$Cd(IO_3)_2$	2.3×10^{-8}

Solid	K_{sp}
$Cd(OH)_2$	4.5×10^{-15}
CdS	1×10^{-27}
$Co(OH)_2$	1×10^{-15}
CoS (α)	5×10^{-22}
(β)	3×10^{-26}
$Cr(OH)_2$	1×10^{-17}
$Cr(OH)_3$	2×10^{-30}
CuBr	5×10^{-9}
CuCl	1.9×10^{-7}
CuI	1×10^{-12}
$Cu(OH)_2$	4.8×10^{-20}
CuS	8×10^{-37}
CuSCN	4.0×10^{-14}
$Fe(OH)_2$	4.1×10^{-15}
$Fe(OH)_3$	2×10^{-39}
$Fe_3(PO_4)_2$	1×10^{-36}
$FePO_4$	4×10^{-27}
FeS	8×10^{-19}
$Ga(OH)_3$	1×10^{-37}
Hg_2Br_2	5.6×10^{-23}
Hg_2Cl_2	1.2×10^{-18}
Hg_2I_2	4.7×10^{-29}

23 SOLUBILITY PRODUCTS *(continued)*

Solid	K_{sp}
HgS (black)	2×10^{-53}
(red)	5×10^{-54}
Hg_2SO_4	7.4×10^{-7}
$MgCO_3$	2×10^{-5}
$MgNH_4PO_4$	3×10^{-13}
$Mg(OH)_2$	7.1×10^{-12}
$Mn(OH)_2$	2×10^{-13}
MnS (pink)	3×10^{-11}
(green)	3×10^{-14}
$Ni(OH)_2$	6×10^{-16}
NiS (α)	4×10^{-20}
(β)	1×10^{-25}
(γ)	3×10^{-27}
$PbBr_2$	2.1×10^{-6}
$PbCO_3$	7.4×10^{-14}
$PbCl_2$	1.7×10^{-5}
$PbCrO_4$	2.5×10^{-13}

Solid	K_{sp}
PbI_2	8×10^{-9}
$Pb(OH)_2$	8×10^{-17}
PbS	3×10^{-28}
$PbSO_4$	1.6×10^{-8}
$Sc(OH)_3$	2×10^{-30}
SnS	1×10^{-26}
$SrCO_3$	5.4×10^{-10}
$SrCrO_4$	2×10^{-5}
$Sr(OH)_2$	3×10^{-4}
$SrSO_4$	3.2×10^{-7}
$V(OH)_2$	4×10^{-16}
$V(OH)_3$	4×10^{-35}
$VO(OH)_2$	3×10^{-24}
$Zn(OH)_2$	3.8×10^{-17}
ZnS (zinc blende)	2×10^{-25}
(wurtzite)	3×10^{-23}

24 SOME IMPORTANT ORGANIC FUNCTIONAL GROUPS

	Functional group[a]	Example	Name
acid anhydride	$R-\mathbf{C(=O)-O-C(=O)}-R$	CH_3COCCH_3 (with C=O at each C)	ethanoic anhydride (acetic anhydride)
acid chloride	$R-\mathbf{C(=O)-Cl}$	CH_3CCl (with C=O)	ethanoyl chloride (acetyl chloride)
alcohol	$R-\mathbf{OH}$	CH_3CH_2OH	ethanol (ethyl alcohol)
aldehyde	$R-\mathbf{C(=O)-H}$	CH_3CH (with C=O)	ethanal (acetaldehyde)
alkene	$R_2\mathbf{C{=}C}R_2$	$CH_2=CH_2$	ethene (ethylene)
alkyne	$R-\mathbf{C{\equiv}C}-R$	$HC\equiv CH$	ethyne (acetylene)
amide	$R-\mathbf{C(=O)-N}(R)-R$	CH_3CNH_2 (with C=O)	ethanamide (acetamide)
amine, primary	$R-\mathbf{NH_2}$	$CH_3CH_2NH_2$	ethylamine
amine, secondary	$R-\mathbf{NH}-R$	$(CH_3CH_2)_2NH$	diethylamine
amine, tertiary	$R-\mathbf{N}(R)-R$	$(CH_3CH_2)_3N$	triethylamine
arene	benzene ring with six R substituents	benzene ring	benzene

[a] Functional group shown in **bold**. R = hydrogen, alkyl or aryl group, except for amines, alcohols, haloalkanes, ketones and thiols where R cannot be H.

24 SOME IMPORTANT ORGANIC FUNCTIONAL GROUPS *(continued)*

	Functional group[a]	Example	Name
carboxylic acid	$R-C(=O)-O-H$	CH_3COOH (O above C)	ethanoic acid (acetic acid)
disulfide	$R-S-S-R$	CH_3SSCH_3	dimethyl disulfide
ester	$R-C(=O)-O-CR_2-R$	CH_3COOCH_3 (O above C)	methyl ethanoate (methyl acetate)
haloalkane	$R-X$	CH_3CH_2Cl	chloroethane (ethyl chloride)
ketone	$R-C(=O)-R$	CH_3COCH_3 (O above C)	propanone (acetone)
phenol	C_6R_5-OH	C_6H_5-OH	phenol
sulfide	$R-S-R$	CH_3SCH_3	dimethyl sulfide
thiol	$R-S-H$	CH_3CH_2SH	ethanethiol(ethyl mercaptan)

[a] Functional group shown in **bold**. R = hydrogen, alkyl or aryl group, except for amines, alcohols, haloalkanes, ketones and thiols where R cannot be H.

25 PROPERTIES OF ORGANIC COMPOUNDS

Note that all compounds listed in the following table should be assumed to be hazardous. Appropriate safety information is available in the relevant MSDS, which may be obtained from www.sigmaaldrich.com/safety-center.html or other online sources.

Table 25 contains names, formulae, physical properties, dipole moments, thermochemical data, and acid dissociation constants of selected organic compounds.

Name International Union of Pure and Applied Chemistry systematic nomenclature is used except where the common (or trivial) name is normally preferred by IUPAC. In some cases alternative names are given in parentheses.

Formula Compounds are depicted using condensed structural formulae, in which parentheses are used in two ways:

(i) for repeating CH_2 groups in straight chain structures, for example, $CH_3(CH_2)_3CH_3$ represents $CH_3CH_2CH_2CH_2CH_3$;

(ii) for side chain groups following the carbon atom of the main chain to which they are attached, for example,

$CH_3CH(CH_3)(CH_2)_2CH_3$ represents $\begin{array}{l} CH_3\underset{\displaystyle |\atop \displaystyle CH_3}{CH}CH_2CH_2CH_3 \end{array}$

$CH_3C(CH_3)(NH_2)CH_3$ represents $CH_3 - \overset{\displaystyle CH_3 \atop \displaystyle |}{\underset{\displaystyle | \atop \displaystyle NH_2}{C}} - CH_3$

Cyclic compounds are depicted using a line between the bonded atoms. For example, cyclobutane, $\begin{array}{ccc} CH_2 & - & CH_2 \\ | & & | \\ CH_2 & - & CH_2 \end{array}$, is depicted as $\underbrace{CH_2(CH_2)_2\,CH_2}$

M (in g mol^{-1}) = molar mass for the formula shown.

ρ (in g cm^{-3}) = density of the solid or liquid form at 298 K and 101.325 kPa, unless otherwise specified. Reported densities followed by (l) refer to the liquid form at a pressure greater than 101.325 kPa at 298 K. Reported densities marked * refer to the liquid at a pressure greater than 101.325 kPa at the specified temperature.

t_m (in °C) = melting temperature.

d = decomposes; s = sublimes; * = melts under pressure; values in parentheses are estimated.

t_b (in °C) = boiling temperature (at 101.325 KPa).

t_{FP} (in °C) = flash point; the lowest temperature at which the application of a small flame causes the vapour above the liquid in a closed cup to ignite.

n = refractive index at 298 K unless otherwise indicated.

p = electric dipole moment for molecules in the gas phase unless otherwise indicated ((l) = liquid, (b) = solution in benzene). Values are given in 10^{-30} C m; 1 debye (D) = 3.336×10^{-30} C m.

All thermochemical data are for the standard state pressure of 10^5 Pa and a temperature of 298 K.

$\Delta_f H^\ominus$ (in kJ mol^{-1}) = the standard enthalpy of formation for 1 mole of the substance, in the gas phase unless otherwise indicated, under standard state conditions, from its elements in their standard reference states (stable forms).

$\Delta_f G^\ominus$ (in kJ mol^{-1}) = the standard Gibbs energy of formation for 1 mole of the substance, in the gas phase unless otherwise indicated, under standard state conditions, from its elements in their standard reference states (stable forms).

$S^\ominus$ (in J K^{-1} mol^{-1}) = the standard entropy of the substance in the gas phase unless otherwise indicated.

$C_p^\ominus$ (in J K^{-1} mol^{-1}) = the standard molar heat capacity at constant pressure of the substance in the gas phase unless otherwise indicated.

$\Delta_{vap}H$ (in kJ mol^{-1}) = the molar enthalpy of vaporisation at 298 K for the transition liquid to vapour.

$\Delta_{sub}H$ (in kJ mol^{-1}) = the molar enthalpy of sublimation at 298 K for the transition solid to vapour.

$\Delta_cH^⊖$ (in kJ mol^{-1}) = the standard molar enthalpy of combustion, for the substance in its stable form at 10^5 Pa and 298 K unless otherwise indicated, the final products being CO_2(g), H_2O(l), N_2(g) and H_2SO_4(aq, 1 mol H_2SO_4 in 115 mol H_2O). In calculating an enthalpy of formation from an enthalpy of combustion, or vice versa, the following standard enthalpies of formation should be used: CO_2(g), −393.5 kJ mol^{-1}; H_2O(l), −285.8 kJ mol^{-1}; H_2SO_4(aq, in 115 mol H_2O), −887.8 kJ mol^{-1}.

$pK_a = -\log K_a$; K_a = −log of the dissociation constant of the acid at 298 K. The dissociation constant of a base B is given in terms of the pK_a value of its conjugate acid BH^+

$pK_{a,1}$ = −log of the first K_a

$pK_{a,2}$ = −log of the second K_a

25 PROPERTIES OF ORGANIC COMPOUNDS

Alkanes

Name	Formula	M g mol^{-1}	ρ g cm^{-3}	t_m °C	t_b °C	t_{FP} °C	n	p 10^{-30} C m (gas)	$\Delta_fH^⊖$ kJ mol^{-1} (gas)	$\Delta_fG^⊖$ kJ mol^{-1} (gas)	$S^⊖$ J K^{-1} mol^{-1} (gas)	$C_p^⊖$ J K^{-1} mol^{-1} (gas)	$\Delta_{vap}H$ kJ mol^{-1}	$\Delta_cH^⊖$ kJ mol^{-1} (state at 298 K)	pK_a
Alkanes															
methane	CH_4	16.0	0.423$^{-162°C}$	−182.5	−161.5			0	−74	−50	186	36	8	−890	
ethane	CH_3CH_3	30.1	0.545$^{-89°C}$	−182.8	−88.6	−135		0	−84	−32	230	53	15	−1560	
propane	$CH_3CH_2CH_3$	44.1	0.493(l)	−187.7	−42.1	−104		0.3	−105	−24	270	74	19	−2220	
butane	$CH_3(CH_2)_2CH_3$	58.1	0.573(l)	−138.3	−0.5	−60	1.348(l)	0	−126	−17	310	97	21	−2877	
pentane	$CH_3(CH_2)_3CH_3$	72.1	0.621	−129.7	36.1	−49	1.355	0	−147	−9	349	120	27	−3509	
hexane	$CH_3(CH_2)_4CH_3$	86.2	0.655	−95.3	68.7	−22	1.372	0	−167	−0.3	388	143	32	−4163	
heptane	$CH_3(CH_2)_5CH_3$	100.2	0.680	−90.6	98.4	−4	1.385	0	−188	8	428	166	37	−4817	
octane	$CH_3(CH_2)_6CH_3$	114.2	0.698	−56.8	125.7	8	1.395	0	−209	16	467	189	42	−5470	
nonane	$CH_3(CH_2)_7CH_3$	128.3	0.714	−53.5	150.8	31	1.403	0	−228	26	506	212	46	−6125	
decane	$CH_3(CH_2)_8CH_3$	142.3	0.726	−29.7	174.1	46	1.410	0	−250	33	545	235	51	−6778	
undecane	$CH_3(CH_2)_9CH_3$	156.3	0.737	−25.6	195.9	61	1.415	0	−271	42	584	257	56	−7431	
dodecane	$CH_3(CH_2)_{10}CH_3$	170.3	0.745	−9.6	216.3	71	1.419	0	−290	52	622	280	61	−8087	
hexadecane	$CH_3(CH_2)_{14}CH_3$	226.4	0.770	18.2	286.8	135	1.433	0	−375	83	778	372	81	−10699	
icosane (eicosane)	$CH_3(CH_2)_{18}CH_3$	282.5	0.785	36.4	343.8	>112	1.441	0	−456	117	934	463	101		
2-methylpropane (isobutene)	$CH_3CH(CH_3)CH_3$	58.1	0.551(l)	−159.6	−11.7	−87	1.352(l)	0.4	−134	−21	295	97	19	−2868	

Name	Formula	$\frac{M}{\text{g mol}^{-1}}$	$\frac{\rho}{\text{g cm}^{-3}}$	$\frac{t_m}{°C}$	$\frac{t_b}{°C}$	$\frac{t_{FP}}{°C}$	n	$\frac{p}{10^{-30}\text{C m}}$ (gas)	$\frac{\Delta_f H^\ominus}{\text{kJ mol}^{-1}}$ (gas)	$\frac{\Delta_f G^\ominus}{\text{kJ mol}^{-1}}$ (gas)	$\frac{S^\ominus}{\text{J K}^{-1}\text{mol}^{-1}}$ (gas)	$\frac{C_p^\ominus}{\text{J K}^{-1}\text{mol}^{-1}}$ (gas)	$\frac{\Delta_{vap}H}{\text{kJ mol}^{-1}}$	$\frac{\Delta_c H^\ominus}{\text{kJ mol}^{-1}}$ (state at 298 K)	pK_a
Alkanes (*cont.*)															
2-methylbutane (isopentane)	$CH_3CH(CH_3)CH_2CH_3$	72.1	0.615	−159.9	27.9	−56	1.351	0.4	−154	−15	344	119	25	−3504	
2,2-dimethylpropane (neopentane)	$CH_3C(CH_3)_2CH_3$	72.1	0.585(l)	−16.6	9.5	−65	1.348(l)	0	−168	−17	306	122	22	−3514	
2-methylpentane (isohexane)	$CH_3CH(CH_3)(CH_2)_2CH_3$	86.2	0.650	−153.7	60.3	−23	1.369	0	−175	−5	381	144	30	−4157	
3-methylpentane	$CH_3CH_2CH(CH_3)CH_2CH_3$	86.2	0.660	−162.9	63.3	−6	1.374	0	−172	−2	380	143	30	−4160	
2,2-dimethylbutane	$CH_3C(CH_3)_2CH_2CH_3$	86.2	0.644	−99.9	49.7	−40	1.366	0	−186	−10	358	142	28	−4148	
2,3-dimethylbutane	$CH_3CH(CH_3)CH(CH_3)CH_3$	86.2	0.657	−128.5	58.0	−33	1.372	0	−178	−4	366	141	29	−4154	
cyclopropane	$CH_2CH_2CH_2$ (ring)	42.1	0.617(l)	−127.4	−32.8	−95	1.380(l)	0	53	104	237	56	20	−2091	
cyclobutane	$CH_2(CH_2)_2CH_2$ (ring)	56.1	$0.704^{0\,°C}$	−90.7	12.5	−65	1.375(l)	0	28	112	265	72	25	−2746	
cyclopentane	$CH_2(CH_2)_3CH_2$ (ring)	70.1	0.740	−93.9	49.3	−37	1.404	0	−76	40	293	83	29	−3291	
cyclohexane	$CH_2(CH_2)_4CH_2$ (ring)	84.2	0.774	6.6	80.7	−18	1.424	0	−123	32	298	106	33	−3920	
Alkenes															
ethene (ethylene)	$CH_2{=}CH_2$	28.1	$0.568^{-104\,°C}$	−169.1	−103.7	−136		0	52	68	220	44	14	−1411	
propene (propylene)	$CH_3CH{=}CH_2$	42.1	0.505(l)	−185.2	−47.7	−108	1.357(l)	1.2	20	62	267	64	18	−2058	
but-1-ene	$CH_3CH_2CH{=}CH_2$	56.1	0.588(l)	−185.3	−6.3	−80	1.396(l)	1.1	0.1	71	306	86	21	−2718	
(2*Z*)-but-2-ene (*cis*-but-2-ene)	$CH_3CH{=}CHCH_3$	56.1	0.616(l)	−138.9	3.7	−73	1.376(l)	0.8	−7	66	301	79	23	−2710	
(2*E*)-but-2-ene (*trans*-but-2-ene)	$CH_3CH{=}CHCH_3$	56.1	0.599(l)	−105.6	0.9	−73	1.367(l)	0	−11	63	297	88	22	−2706	
pent-1-ene	$CH_3(CH_2)_2CH{=}CH_2$	70.1	0.635	−165.2	30.0	−28	1.368	1.7	−21	79	346	110	26	−3350	

Name	Formula	M / g mol^{-1}	ρ / g cm^{-3}	t_m / °C	t_b / °C	t_{FP} / °C	n	p / 10^{-30} C m (gas)	$\Delta_f H^\ominus$ / kJ mol^{-1} (gas)	$\Delta_f G^\ominus$ / kJ mol^{-1} (gas)	$S^\ominus$ / J K^{-1} mol^{-1} (gas)	$C_p^\ominus$ / J K^{-1} mol^{-1} (gas)	$\Delta_{vap}H$ / kJ mol^{-1}	$\Delta_c H^\ominus$ / kJ mol^{-1} (state at 298 K)	pK_a
Alkenes (*cont.*)															
(2*Z*)-pent-2-ene (*cis*-pent-2-ene)	$CH_3CH_2CH=CHCH_3$	70.1	0.650	−151.4	36.9	−27	1.380	0	−28	72	346	102	26	−3343	
(2*E*)-pent-2-ene (*trans*-pent-2-ene)	$CH_3CH_2CH=CHCH_3$	70.1	0.643	−140.2	36.4	−45	1.376	0	−32	70	340	108	26	−3338	
hex-1-ene	$CH_3(CH_2)_3CH=CH_2$	84.2	0.668	−139.8	63.5	−20	1.385	0	−43	86	385	132	31	−4002	
(2*Z*)-hex-2-ene (*cis*-hex-2-ene)	$CH_3(CH_2)_2CH=CHCH_3$	84.2	0.683	−141.1	68.8		1.395	0	−52	76	387	126	32	−3992	
(2*E*)-hex-2-ene (*trans*-hex-2-ene)	$CH_3(CH_2)_2CH=CHCH_3$	84.2	0.673	−133.0	67.9		1.391	0	−54	76	381	132	32	−3991	
(3*Z*)-hex-3-ene (*cis*-hex-3-ene)	$CH_3CH_2CH=CHCH_2CH_3$	84.2	0.675	−137.8	66.4		1.392	0	−48	83	380	124	31	−3997	
(3*E*)-hex-3-ene (*trans*-hex-3-ene)	$CH_3CH_2CH=CHCH_2CH_3$	84.2	0.672	−115.4	67.1		1.391	0	−54	78	375	133	32	−3990	
hept-1-ene	$CH_3(CH_2)_4CH=CH_2$	98.2	0.693	−119.0	93.6	−1	1.397	0	−62	96	424	155	36	−4658	
oct-1-ene	$CH_3(CH_2)_5CH=CH_2$	112.2	0.711	−101.7	121.3	21	1.406	0	−81	106	463	178	41	−5313	
2-methylpropene (isobutylene)	$CH_3C(CH_3)=CH_2$	56.1	0.58(l)	−140.4	−6.9		1.393$^{-25°C}$	1.7	−17	58	294	89	21	−2700	
2-methylbut-1-ene	$CH_3CH_2C(CH_3)=CH_2$	70.1	0.645	−137.6	31.2	−34	1.375	1.7(l)	−35	67	340	112	26	−3336	
3-methylbut-1-ene	$CH_3CH(CH_3)CH=CH_2$	70.1	0.621(l)	−168.5	20.1	−56	1.364(l)	0.8	−28	76	334	119	24	−3369	
2-methylbut-2-ene	$CH_3CH=C(CH_3)CH_3$	70.1	0.657	−133.8	38.6	−45	1.384	1.7(b)	−42	60	339	105	27	−3328	
2,3-dimethylbut-2-ene	$CH_3C(CH_3)=C(CH_3)CH_3$	84.2	0.703	−74.3	73.2	−16	1.409		−68	67	365	128	33	−3974	
cyclopropene	$CH_2CH=CH$ (ring)	40.1			d–36			1.5	277	287	244			−2029(g)	
cyclobutene	$CH_2CH_2CH=CH$ (ring)	54.1	0.733$^{0°C}$		2			0.4	157	202	264	67		−2588	
cyclopentene	$CH_2(CH_2)_2CH=CH$ (ring)	68.1	0.767	−135.1	44.2	−34	1.419	0.7	34	112	290	75	29	−3115	

Name	Formula	M / g mol^{-1}	ρ / g cm^{-3}	t_m / °C	t_b / °C	t_{FP} / °C	n	p / 10^{-30} C m (gas)	$\Delta_f H^\ominus$ / kJ mol^{-1} (gas)	$\Delta_f G^\ominus$ / kJ mol^{-1} (gas)	$S^\ominus$ / J K^{-1} mol^{-1} (gas)	$C_p^\ominus$ / J K^{-1} mol^{-1} (gas)	$\Delta_{vap}H$ / kJ mol^{-1}	$\Delta_c H^\ominus$ / kJ mol^{-1} (state at 298 K)	pK_a
Alkenes (*cont.*)															
cyclohexene	$\overline{CH_2(CH_2)_3CH=CH}$ (ring)	82.1	0.806	−103.5	83.0	−12	1.444	1.1	−5	107	311	105	34	−3752	
propadiene (allene)	$CH_2=C=CH_2$	40.1	0.584(1)	−136.2	−34.5		1.417	0	191	201	244	59		−1943	
buta-1,2-diene	$CH_3CH=C=CH_2$	54.1	$0.676^{0\,°C}$	−136.2	10.9		1.420(l)	1.3	162	198	293	80	23	−2594	
buta-1,3-diene	$CH_2=CHCH=CH_2$	54.1	0.615(1)	−108.9	−4.4	−76	1.416	0	110	151	279	80	22	−2542	
2-methylbuta-1,3-diene (isoprene)	$CH_2=CHC(CH_3)=CH_2$	68.1	0.676	−146.0	34.1	−53	1.419	0.9	76	146	316	105	27	−3159	
cyclopenta-1,3-diene	$\overline{CH_2CH=CHCH=CH}$ (ring)	66.1	$0.802^{20\,°C}$	−85	41	−3	$1.444^{20\,°C}$	1.4	134	179	268		28	−2931	
cyclohexa-1, 3-diene	$\overline{CH_2CH_2CH=CHCH=CH}$ (ring)	80.1	$0.841^{20\,°C}$	−89	80.5		$1.476^{20\,°C}$	1.5	106				33		
Alkynes															
ethyne (acetylene)	$CH\equiv CH$	26.0	0.377	−80.8	s−84.0			0	228	211	201	44	18	−1301	
propyne	$CH_3C\equiv CH$	40.1	0.607	−102.7	−23.2		1.386(l)	2.6	185	194	248	61		−1937	
but-1-yne	$CH_3CH_2C\equiv CH$	54.1	$0.678^{0\,°C}$	−125.7	8.0		−31.396(l)	2.7	165	202	291	81	23	−2597	
but-2-yne	$CH_3C\equiv CCH_3$	54.1	0.686	−32.3	27.0		1.389		146	185	283	78	27	−2551	
pent-1-yne	$CH_3(CH_2)_2CCH$	68.1	$0.690^{20\,°C}$	−105.7	40.2	−34	1.382	2.7	144	210	330	107			
pent-2-yne	$CH_3CH_2C\equiv CCH_3$	68.1	0.706	−109.3	56.1	−20	1.401		129	194	332	99			
hex-1-yne	$CH_3(CH_2)_3C\equiv CH$	82.1	0.710	−131.9	71.3	−21	1.396	2.8	124	219	369	128			
Aromatic hydrocarbons															
benzene	C_6H_6	78.1	0.874	5.5	80.1	−11	1.498	0	83	130	269	82	34	−3268	
methylbenzene (toluene)	$C_6H_5CH_3$	92.1	0.862	−95.0	110.6	4	1.494	1.2	50	122	321	104	38	−3910	
ethylbenzene	$C_6H_5CH_2CH_3$	106.2	0.863	−95.0	136.2	15	1.493	2.0	30	131	361	128	42	−4565	
propylbenzene	$C_6H_5CH_2CH_2CH_3$	120.2	0.858	−99.5	159.2	47	1.490	1.2(b)	8	137	401	152	46	−5218	

Name	Formula	$\frac{M}{\text{g mol}^{-1}}$	$\frac{\rho}{\text{g cm}^{-3}}$	$\frac{t_m}{°\text{C}}$	$\frac{t_b}{°\text{C}}$	$\frac{t_{FP}}{°\text{C}}$	n	$\frac{p}{10^{-30}\text{C m}}$ (gas)	$\frac{\Delta_f H^\ominus}{\text{kJ mol}^{-1}}$ (gas)	$\frac{\Delta_f G^\ominus}{\text{kJ mol}^{-1}}$ (gas)	$\frac{S^\ominus}{\text{J K}^{-1}\text{ mol}^{-1}}$ (gas)	$\frac{C_p^\ominus}{\text{J K}^{-1}\text{ mol}^{-1}}$ (gas)	$\frac{\Delta_{vap}H}{\text{kJ mol}^{-1}}$	$\frac{\Delta_c H^\ominus}{\text{kJ mol}^{-1}}$ (state at 298 K)	pK_a
Aromatic hydrocarbons *(cont.)*															
isopropylbenzene (cumene)	$C_6H_5CH(CH_3)CH_3$	120.2	0.858	−96.0	152.4	34	1.489	2.6	4	137	389	152	45	−5216	
2-phenyl-2-methylpropane (*tert*-butylbenzene)	$C_6H_5C(CH_3)_2CH_3$	134.2	0.862	−57.9	169.1	44	1.490	2.8	−71(l)	136(l)	279(l)		48	−5865	
1,2-dimethylbenzene (*o*-xylene)	$C_6H_4(CH_3)_2$	106.2	0.876	−25.2	144.4	32	1.503	2.1	19	122	353	133	43	−4553	
1,3-dimethylbenzene (*m*-xylene)	$C_6H_4(CH_3)_2$	106.2	0.860	−47.9	139.1	25	1.495	1.2(b)	17	119	358	128	43	−4552	
1,4-dimethylbenzene (*p*-xylene)	$C_6H_4(CH_3)_2$	106.2	0.857	13.3	138.4	26	1.493	0	18	121	352	127	42	−4553	
1,3,5-trimethylbenzene (mesitylene)	$C_6H_3(CH_3)_3$	120.2	0.861	−44.7	164.7	44	1.497	0	−16	118	385	150	47	−5193	
1,2,3,4-tetramethylbenzene	$C_6H_2(CH_3)_4$	134.2	0.902	−6.2	205.0	68	1.518		−42	123	417	190	48	−5846	
naphthalene	$C_{10}H_8$	128.2	$1.025^{20\,°\text{C}}$	80.3	217.9	78	$1.582^{99\,°\text{C}}$	0	150	223	336	133	72(sub)	−5156	
anthracene	$C_{14}H_{10}$	178.2	1.28	216	340	121		0(b)	129(s)	286(s)	207(s)	211(s)	102(sub)	−7068	
phenanthrene	$C_{14}H_{10}$	178.2	1.179	99.2	340		1.594	0(b)	116(s)	272(s)	212(s)		91(sub)	−7054	
Halogen-containing compounds															
bromomethane (methyl bromide)	CH_3Br	94.9	$1.676^{20\,°\text{C}*}$	−93.7	3.6		1.422(l)	6.0	−36	−26	246	42	23		
dibromomethane	CH_2Br_2	173.8	2.484	−52.7	97.0		1.539	4.8	−4	−6	293	55			
tribromomethane (bromoform)	$CHBr_3$	252.7	2.878	8.1	149.6		1.596	3.3	17	8	331	71	45		
tetrabromomethane	CBr_4	331.6	3.420	92	189.5		1.594(l)	0	79	67	358	91	60(sub)		
chloromethane (methyl choride)	CH_3Cl	50.5	0.911(l)	−97.7	−24.2		1.371(l)	6.2	−82	−58	235	41	22		
dichloromethane	CH_2Cl_2	84.9	1.319	−95.1	39.8		1.421	5.3	−95	−69	270	51	29		

Name	Formula	M / g mol^{-1}	ρ / g cm^{-3}	t_m / °C	t_b / °C	t_{FP} / °C	n	p / 10^{-30} C m (gas)	$\Delta_f H^\ominus$ / kJ mol^{-1} (gas)	$\Delta_f G^\ominus$ / kJ mol^{-1} (gas)	$S^\ominus$ / J K^{-1} mol^{-1} (gas)	$C_p^\ominus$ / J K^{-1} mol^{-1} (gas)	$\Delta_{vap}H$ / kJ mol^{-1}	$\Delta_c H^\ominus$ / kJ mol^{-1} (state at 298 K)	pK_a
Halogen-containing compounds (*cont.*)															
trichloromethane (chloroform)	$CHCl_3$	119.4	1.480	−63.5	61.1		1.443	3.4	−103	−70	296	66	31		
tetrachloromethane (carbon tetrachloride)	CCl_4	153.8	1.584	−23	76.7		1.457	0	−96	−54	310	83	32		
fluoromethane (methyl fluoride)	CH_3F	34.0	0.557(l)	−141.8	−78.4		1.167(l)	6.2	−234	−210	223	38			
difluoromethane	CH_2F_2	52.0	$1.214^{-52\,°C}$	−136.0	−51.6		1.16(l)	6.6	−453	−425	247	43			
trifluoromethane (fluoroform)	CHF_3	70.0	0.673	−155.1	−82.1			5.5	−695	−661	260	51	17		
tetrafluoromethane	CF_4	88.0	3.034	−183.5	−128.0			0	−933	−888	262	61			
iodomethane (methyl iodide)	CH_3I	141.9	2.265	−66.5	42.4		1.527	5.4	15	16	254	44	27		
diiodomethane	CH_2I_2	267.8	3.308	6.1	182	>110	1.738	3.6(b)	113	96	310	58	46		
triiodomethane (iodoform)	CHI_3	393.7	4.008	119	218			2.9(b)	211	178	356	75	70(sub)		
tetraiodomethane	CI_4	519.6	$4.23^{20\,°C}$	171	d			0			392	96			
bromo(chloro)-difluoromethane	$CBrClF_2$	165.4		−159.5	−3.7				−472	−448	319	75			
bromo(difluoro)methane	$CHBrF_2$	130.9	$1.55^{16\,°C*}$	−145.0	−14.6			5.0	−464	−447	295	59			
dibromo(difluoro)methane	CBr_2F_2	209.8	$2.29^{15\,°C}$	−142.0	24.5		$1.402^{20\,°C}$	2.2	−430	−419	325	77			
chloro(fluoro)methane (Freon 31)	CH_2ClF	68.5		−133.0	−9.1			6.1	−291	−266	264	47			
chloro(difluoro)methane (Freon 22 or CFC-22)	$CHClF_2$	86.5	$1.491^{-69\,°C}$	−157.4	−40.7			4.7	−483	−451	281	56			
chloro(trifluoro)methane (Freon 13)	$CClF_3$	104.5		−181	−81.4			1.7	−706	−666	286	67			
dichloro(fluoro)methane (Freon 21)	$CHCl_2F$	102.9	$1.405^{9\,°C}$	−135	8.9		1.372(l)	4.3	−299	−268	293	61			

Name	Formula	M / g mol^{-1}	ρ / g cm^{-3}	t_m / °C	t_b / °C	t_{FP} / °C	n	p / 10^{-30} C m (gas)	$\Delta_f H^\ominus$ / kJ mol^{-1} (gas)	$\Delta_f G^\ominus$ / kJ mol^{-1} (gas)	$S^\ominus$ / J K^{-1} mol^{-1} (gas)	$C_p^\ominus$ / J K^{-1} mol^{-1} (gas)	$\Delta_{vap}H$ / kJ mol^{-1}	$\Delta_c H^\ominus$ / kJ mol^{-1} (state at 298 K)	pK_a
Halogen-containing compounds (*cont.*)															
dichlorodifluoromethane (Freon 12)	CCl_2F_2	120.9	1.49(l)	−158.0	−29.8			1.7	−477	−439	301	72			
trichloro(fluoro)methane (Freon 11)	CCl_3F	137.4	1.488$^{20 °C}$	−111.1	23.7		1.384(l)	1.5	−276	−238	310	78	25		
bromoethane (ethyl bromide)	CH_3CH_2Br	109.0	1.451	−118.6	38.4	−23	1.421	6.8	−62	−24	287	65	28		
1,2-dibromoethane	CH_2BrCH_2Br	187.9	2.170	10.0	131.7		1.536	3.7	−39	−11	330	85	40		
chloroethane (ethyl chloride)	CH_3CH_2Cl	64.5	0.924$^{0 °C}$	−136.4	12.3	−43	1.365(l)	6.8	−112	−60	276	63	24		
1,1-dichloroethane	CH_3CHCl_2	99.0	1.168	−97.0	57.3	<−18	1.414	6.9	−128	−71	305	76	31		
1,2-dichloroethane	CH_2ClCH_2Cl	99.0	1.246	−35.7	83.5	13	1.442	4.8	−127	−71	308	79	40		
1,1,2-trichloroethane	$CH_2ClCHCl_2$	133.4	1.432	−36.6	113.8		1.469	4.7	−151	−90	337	89	40		
1,1,2,2-tetrachloroethane	$CHCl_2CHCl_2$	167.9	1.588	−43.8	146.2		1.491	4.4	−149	−82	363	101	46		
pentachloroethane	$CHCl_2CCl_3$	202.3	1.674	−29	162		1.501	3.1	−142	−67	381	118	46		
hexachloroethane	CCl_3CCl_3	236.7	2.091$^{20 °C}$	*187	s			0(b)	−141	−57	397	136	65(sub)		
1,1,1,2-tetrafluoroethane (HFC−134a)	CH_2FCF_3	102.0	1.207(l)	−103.3	−26.1			6.0			317	87	22		
iodoethane	CH_3CH_2I	156.0	1.925	−111.1	72.3	53	1.510	6.4	−8	19	306	67	32		
1-bromopropane	$CH_3CH_2CH_2Br$	123.0	1.345	−110.0	71.0	25	1.432	7.3	−88	−23	331	86	35		
2-bromopropane	$CH_3CHBrCH_3$	123.0	1.302	−89.0	59.4	19	1.422	7.4	−99	−29	316	89	31		
1-chloropropane	$CH_3CH_2CH_2Cl$	78.5	0.890$^{20 °C}$	−122.8	46.6	−18	1.386	6.8	−132	−53	319	85	29		
2-chloropropane	$CH_3CHClCH_3$	78.5	0.855	−117.2	35.7	−35	1.375	7.2	−145	−61	304	87	27		
1-iodopropane	$CH_3CH_2CH_2I$	170.0	1.739	−101.3	102.5	34	1.503	6.8	−31	28	336	90	36		
1-bromobutane	$CH_3(CH_2)_3Br$	137.0	1.269	−112.4	101.6	23	1.438	6.9	−107	−13	370	109	37		
2-bromobutane	$CH_3CH_2CHBrCH_3$	137.0	1.253	−111.9	91.2	21	1.434	7.4	−120	−26	370	111	35		

Name	Formula	M / g mol^{-1}	ρ / g cm^{-3}	t_m / °C	t_b / °C	t_{FP} / °C	n	p / 10^{-30} C m (gas)	$\Delta_f H^\ominus$ / kJ mol^{-1} (gas)	$\Delta_f G^\ominus$ / kJ mol^{-1} (gas)	$S^\ominus$ / J K^{-1} mol^{-1} (gas)	$C_p^\ominus$ / J K^{-1} mol^{-1} (gas)	$\Delta_{vap}H$ / kJ mol^{-1}	$\Delta_c H^\ominus$ / kJ mol^{-1} (state at 298 K)	pK_a
Halogen-containing compounds (*cont.*)															
1-chlorobutane	$CH_3(CH_2)_3Cl$	92.6	0.881	−123.1	78.4	−6	1.400	6.8	−155	−46	358	108	34		
2-chlorobutane	$CH_3CH_2CHClCH_3$	92.6	0.868	−131.3	68.3	−15	1.394	6.8	−162	−54	360	108	32		
1-chloro-2-methylpropane	$CH_3CH(CH_3)CH_2Cl$	92.6	0.872	−130.3	68.9	−10	1.395	6.7	−159	−50	354	108	32		
2-chloro-2-methylpropane	$CH_3CCl(CH_3)CH_3$	92.6	0.836	−25.4	50.7	−10	1.383	7.1	−183	−64	322	114	29		
1-iodobutane	$CH_3(CH_2)_3I$	184.0	1.607	−103.0	130.5	33	1.497	7.1							
1-bromopentane	$CH_3(CH_2)_4Br$	151.1	1.212	−95.0	129.6	31	1.442	7.3	−129	−6	409	132	41		
1-chloropentane	$CH_3(CH_2)_4Cl$	106.6	0.880	−99.0	107.8	11	1.410	7.2	−175	−37	397	130	38		
1-bromohexane	$CH_3(CH_2)_5Br$	165.1	1.169	−84.7	155.3	57	1.445		−148				46		
1-chlorohexane	$CH_3(CH_2)_5Cl$	120.6	0.874	−94.0	134.3	26	1.418	6.5(b)							
bromoethene (vinyl bromide)	$CH_2{=}CHBr$	107.0	$1.493^{20\,°C*}$	−137.8	15.8	−18	1.435(l)	4.7	79	82	276	56			
chloroethene (vinyl chloride)	$CH_2{=}CHCl$	62.5	$0.911^{20\,°C*}$	−153.7	−13.3		1.366(l)	4.8	37	54	264	54	23		
1,1-dichloroethene	$CH_2{=}CCl_2$	96.9	$1.213^{20\,°C}$	−122.5	31.6	−10	$1.425^{20\,°C}$	4.5	2	25	289	67	27		
(*Z*)-1,2-dichloroethene (*cis*)	$CHCl{=}CHCl$	96.9	1.272	−80.5	60.3	6	1.446	6.3	5	28	290	65	31		
(*E*)-1,2-dichloroethene (*trans*)	$CHCl{=}CHCl$	96.9	1.257	−49.8	47.7	6	1.443	0	6	29	290	67	29		
1,1-difluoroethene	$CH_2{=}CF_2$	64.0		−144	−85.7			4.6	−335	−311	265	59			
(*Z*)-1-chloropropene (*cis*)	$CH_3CH{=}CHCl$	76.5	$0.935^{20\,°C}$	−134.8	32.8		1.403	5.6							
(*E*)-1-chloropropene (*trans*)	$CH_3CH{=}CHCl$	76.5	$0.935^{20\,°C}$	−99	37.4		1.403	6.6							
chloroethyne	$CH{\equiv}CCl$	60.5		−126	−30			1.5			242	54			
chlorobenzene	C_6H_5Cl	112.6	1.101	−45.2	131.7	23	1.522	5.6	52	99	314	98	41		
1,2-dichlorobenzene	$C_6H_4Cl_2$	147.0	1.300	−17.0	180.5	65	1.549	8.3	30	83	342	113	48		

Name	Formula	M / g mol^{-1}	ρ / g cm^{-3}	t_m / °C	t_b / °C	t_{FP} / °C	n	p / 10^{-30}C m (gas)	$\Delta_f H^\ominus$ / kJ mol^{-1} (gas)	$\Delta_f G^\ominus$ / kJ mol^{-1} (gas)	$S^\ominus$ / J K^{-1} mol^{-1} (gas)	$C_p^\ominus$ / J K^{-1} mol^{-1} (gas)	$\Delta_{vap}H$ / kJ mol^{-1}	$\Delta_c H^\ominus$ / kJ mol^{-1} (state at 298 K)	pK_a
Halogen-containing compounds (*cont.*)															
1,3-dichlorobenzene	$C_6H_4Cl_2$	147.0	1.283	–24.8	173.0	65	1.543	5.7	26	79	343	114	48		
1,4-dichlorobenzene	$C_6H_4Cl_2$	147.0	1.248	52.7	174	65	1.528$^{20\,°C}$	0	23	77	337	114	65(sub)		
(chloromethyl)benzene (benzyl chloride)	$C_6H_5CH_2Cl$	126.6	1.095	–45.0	179.3	60	1.536	6.1(b)	19			51			
Nitro compounds															
nitromethane	CH_3NO_2	61.0	1.131	–28.6	101.2	35	1.380	11.5	–75	–7	275	57	38	–709	
nitroethane	$C_2H_5NO_2$	75.1	1.045	–89.5	114.1	30	1.390	12.0	–102	–6	315	78	42	–1358	
1-nitropropane	$C_3H_7NO_2$	89.1	0.996	–108	131.2	25	1.400	12.2	–124	1.6	356	102	43	–2014	
2-nitropropane	$C_3H_7NO_2$	89.1	0.982	–91.3	120.2	26	1.394$^{20\,°C}$	12.4	–139	–12	348	102	41	–2001	
1-nitrobutane	$C_4H_9NO_2$	103.1	0.968	–81.3	152.8	47	1.408	12.0	–144	10	394	125	49	–2668	
nitrobenzene	$C_6H_5NO_2$	123.1	1.198	5.7	210.8	87	1.550	14.1	13(l)	143(l)	224(l)	186(l)	55	–3088	
1,2-dinitrobenzene	$NO_2C_6H_4NO_2$	168.1	1.565$^{17\,°C}$	118.5	318			20(b)	–2(s)					–2931	
1,3-dinitrobenzene	$NO_2C_6H_4NO_2$	168.1	1.575$^{18\,°C}$	90.0	291			13(b)	54				81(sub)	–2905	
1,4-dinitrobenzene	$NO_2C_6H_4NO_2$	168.1	1.625$^{18\,°C}$	174	297			0.0	–39(s)					–2894	
2,4,6-trinitrotoluene	$CH_3C_6H_2(NO_2)_3$	227.1	1.654	80.1	d240			4.0(b)	43				105(sub)	–3402	
Alcohols															
methanol (methyl alcohol)	CH_3OH	32.0	0.787	–97.7	64.7	12	1.327	5.7	–201	–163	240	44	38	–726	
ethanol (ethyl alcohol)	CH_3CH_2OH	46.1	0.785	–114.1	78.3	8	1.359	5.6	–235	–168	283	65	43	–1367	
propan-1-ol	$CH_3(CH_2)_2OH$	60.1	0.800	–126.2	97.2	15	1.384	5.6	–255	–160	323	86	47	–2021	
propan-2-ol (isopropyl alcohol)	$CH_3CH(OH)CH_3$	60.1	0.781	–89.5	82.3	12	1.375	5.5	–273	–173	309	89	45	–2006	
butan-1-ol	$CH_3(CH_2)_3OH$	74.1	0.806	–89.8	117.7	30	1.397	5.5	–275	–151	363	110	52	–2676	
butan-2-ol (*sec*-butyl alcohol)	$CH_3CH_2CH(OH)CH_3$	74.1	0.802	–114.7	99.5	23	1.395	5.5(b)	–293	–168	359	113	50	–2661	

Name	Formula	M / g mol^{-1}	ρ / g cm^{-3}	t_m / °C	t_b / °C	t_{FP} / °C	n	p / 10^{-30} C m (gas)	$\Delta_f H^\ominus$ / kJ mol^{-1} (gas)	$\Delta_f G^\ominus$ / kJ mol^{-1} (gas)	$S^\ominus$ / J K^{-1} mol^{-1} (gas)	$C_p^\ominus$ / J K^{-1} mol^{-1} (gas)	$\Delta_{vap} H$ / kJ mol^{-1}	$\Delta_c H^\ominus$ / kJ mol^{-1} (state at 298 K)	pK_a
Alcohols *(cont.)*															
2-methylpropan-1-ol (isobutyl alcohol)	$CH_3CH(CH_3)CH_2OH$	74.1	0.798	−108.0	107.9	29	1.394	5.5	−283			113	51	−2668	
2-methylpropan-2-ol (*tert*-butyl alcohol)	$CH_3C(CH_3)(OH)CH_3$	74.1	0.781	25.7	82.6	11	1.385	5.6(b)	−313	−178	327	114	47	−2644(l)	
pentan-1-ol	$CH_3(CH_2)_4OH$	88.2	0.816	−78.9	138.0	33	1.408	5.7(b)	−295	−142	403	133	57	−3331	
pentan-2-ol	$CH_3(CH_2)_2CH(OH)CH_3$	88.2	0.805	−73.0	119.0	33	1.404	5.5(b)	−313				52	−3317	
pentan-3-ol	$CH_3CH_2CH(OH)CH_2CH_3$	88.2	0.816	−69.0	115.6	33	1.408	5.5(b)	−317				52	−3314	
2-methylbutan-1-ol	$CH_3CH_2CH(CH_3)CH_2OH$	88.2	0.815		128	50	$1.409^{20\,°C}$		−302				54	−3326	
3-methylbutan-1-ol	$CH_3CH(CH_3)CH_2CH_2OH$	88.2	0.805	−117.2	131.1	45	$1.405^{20\,°C}$	6.1(b)	−301				55	−3326	
2-methylbutan-2-ol	$CH_3CH_2C(CH_3)(OH)CH_3$	88.2	0.805	−9.0	102.4	21	1.402	5.7(b)	−331	−167.3	367	132	49	−3303	
3-methylbutan-2-ol	$CH_3CH(CH_3)CH(OH)CH_3$	88.2	0.813		112.9	39	$1.409^{20\,°C}$		−315				52	−3316	
2,2-dimethylpropan-1-ol (neopentyl alcohol)	$CH_3C(CH_3)_2CH_2OH$	88.2	$0.812^{20\,°C}$	52.5	113.0	+36			−399(l)					−3283(l)	
hexan-1-ol	$CH_3(CH_2)_5OH$	102.2	$0.814^{20\,°C}$	−44.6	157.5	60	1.416	5.2(b)	−316	−134	442	156	62	−3984	
heptan-1-ol	$CH_3(CH_2)_6OH$	116.2	$0.822^{20\,°C}$	−34.0	176.2	73	1.422	5.8(b)	−336	−126.3	480	179	67	−4638	
octan-1-ol	$CH_3(CH_2)_7OH$	130.2	0.825	−15.5	195.2	81	1.428	5.3(b)	−356	−116	519	202	71	−5294	
cyclohexanol	$\underline{CH_2(CH_2)_4CH}OH$ (ring)	100.2	$0.962^{20\,°C}$	25.2	161.0	67	1.465	6.0(b)	−286	−109.3	328	127	62	−3728(l)	
phenylmethanol (benzyl alcohol)	$C_6H_5CH_2OH$	108.1	1.042	−15.2	205.3	93	1.538	5.7	−161(l)	−27(l)	217(l)	218(l)	60	−3737	
ethane-1, 2-diol (ethylene glycol)	$CH_2(OH)CH_2OH$	62.1	1.110	−13.0	198	110	1.431	7.6	−388	−304	324	97	68	−1189	
Phenols															
phenol (hydroxybenzene)	C_6H_5OH	94.1	$1.058^{20\,°C}$	40.9	181.8	79	$1.542^{41\,°C}$	4.8	−96	−33	316	104	69(sub)	−3053(s)	9.98
benzene-1,2-diol (pyrocatechol)	OHC_6H_4OH	110.1	$1.344^{20\,°C}$	105	245	137	1.60	8.8(b)	−268				87(sub)	−2858(s)	9.45, 12.8

Name	Formula	M / g mol^{-1}	ρ / g cm^{-3}	t_m / °C	t_b / °C	t_{FP} / °C	n	p / 10^{-30} C m (gas)	$\Delta_f H^\ominus$ / kJ mol^{-1} (gas)	$\Delta_f G^\ominus$ / kJ mol^{-1} (gas)	$S^\ominus$ / J K^{-1} mol^{-1} (gas)	$C_p^\ominus$ / J K^{-1} mol^{-1} (gas)	$\Delta_{vap}H$ / kJ mol^{-1}	$\Delta_c H^\ominus$ / kJ mol^{-1} (state at 298 K)	pK_a
Phenols (*cont.*)															
benzene-1,3-diol (resorcinol)	OHC_6H_4OH	110.1	1.272	110	276	127		7.0(b)	−275				93(sub)	−2851(s)	9.3, 11.1
benzene-1,4-diol (hydroquinone)	OHC_6H_4OH	110.1	$1.328^{15\,°C}$	172.3	287			4.7(b)	−262	−176.3	344	316(s)	99(sub)	−2853(s)	9.9, 11.4
2-aminophenol	$NH_2C_6H_4OH$	109.1	1.328	174											4.70[a], 9.9
3-aminophenol	$NH_2C_6H_4OH$	109.1		123											4.3[a], 9.8
4-aminophenol	$NH_2C_6H_4OH$	109.1		187.5					−179(s)						5.65[a], 10.4
2-chlorophenol	ClC_6H_4OH	128.6	1.257	9.8	174.9	63	$1.552^{20\,°C}$	7.3	−185(l)						8.53
3-chlorophenol	ClC_6H_4OH	128.6	1.218	33	214	>110	$1.556^{40\,°C}$	7.3(b)	−153				53(sub)		9.13
4-chlorophenol	ClC_6H_4OH	128.6	$1.265^{40\,°C}$	42.7	220	115	$1.558^{40\,°C}$	7.0	−146				52(sub)		9.43
2-ethylphenol	$C_2H_5C_6H_4OH$	122.2	1.015	−3.3	204.5	78	$1.537^{20\,°C}$		−145				64	−4368(s)	10.3
3-ethylphenol	$C_2H_5C_6H_4OH$	122.2	1.008	−4.0	218.4	94	$1.533^{20\,°C}$		−146				68	−4363(s)	10.07
4-ethylphenol	$C_2H_5C_6H_4OH$	122.2	1.054	45.1	219	100	1.524	5.3(b)	−144				80(sub)	−4353(s)	10.2
2-methylphenol (*o*-cresol)	$CH_3C_6H_4OH$	108.1	1.135	30.9	191.0	81	1.544	4.8(b)	−129	−37	358	130	76(sub)	−3693(s)	10.26
3-methylphenol (*m*-cresol)	$CH_3C_6H_4OH$	108.1	1.030	12.2	202.2	86	1.540	4.9(b)	−132	−41	357	125	62	−3704(s)	10.99
4-methylphenol (*p*-cresol)	$CH_3C_6H_4OH$	108.1	1.029	34.7	201.9	89	1.531	4.9(b)	−125	−31	348	125	74(sub)	−3699(s)	10.26
2-nitrophenol	$NO_2C_6H_4OH$	139.1	$1.304^{30\,°C}$	45.1	216		$1.572^{50\,°C}$	10.5(b)	−202(s)						7.24
3-nitrophenol	$NO_2C_6H_4OH$	139.1	$1.280^{100\,°C}$	97				13.0(b)	−210(s)						8.38
4-nitrophenol	$NO_2C_6H_4OH$	139.1	$1.479^{20\,°C}$	113.8	d279.0			16.8(b)	−194(s)						7.15
2,4-dinitrophenol	$(NO_2)_2C_6H_3OH$	184.1	$1.683^{24\,°C}$	115	s			10.7(b)	−128				105(sub)		4.08
2,4,6-trinitrophenol (picric acid)	$(NO_2)_3C_6H_2OH$	229.1	$1.763^{20\,°C}$	122.5	d300.0			5.0(b)	−214(s)						0.37

[a] pK_a of the conjugate acid, BH^+.

Name	Formula	M / g mol^{-1}	ρ / g cm^{-3}	t_m / °C	t_b / °C	t_{FP} / °C	n	p / 10^{-30} C m (gas)	$\Delta_f H^\ominus$ / kJ mol^{-1} (gas)	$\Delta_f G^\ominus$ / kJ mol^{-1} (gas)	$S^\ominus$ / J K^{-1} mol^{-1} (gas)	$C_p^\ominus$ / J K^{-1} mol^{-1} (gas)	$\Delta_{vap}H$ / kJ mol^{-1}	$\Delta_c H^\ominus$ / kJ mol^{-1} (state at 298 K)	pK_a
Thiols															
butane-1-thiol	$CH_3(CH_2)_3SH$	90.2	0.837	−116	98.5	1	1.443	5.1(l)	−88	11	375	118	37		
2-methylpropane-2-thiol	$CH_3C(CH_3)(SH)CH_3$	90.2	0.794	−0.5	64.3	−10	1.420	5.5	−110	0.7	338	121	31		11.22
benzenethiol (thiophenol)	C_6H_5SH	110.2	1.073	−14.9	169	50	1.586	3.8(l)	112	149	337	105	49		6.62
Amines															
methanamine (methylamine)	CH_3NH_2	31.1	0.656(l)	−93.5	−6.3		1.349(l)	4.4	−23	32	243	50	24	−1060(l)	10.64[a]
N-methylmethanamine (dimethylamine)	$(CH_3)_2NH$	45.1	0.680$^{0\,°C}$(l)	−92.2	6.9		1.357(l)	3.4	−19	68	273	71	25	−1743(l)	10.87[a]
N,N-dimethylmethanamine (trimethylamine)	$(CH_3)_3N$	59.1	0.627(l)	−117.1	2.9	−6	1.363$^{0\,°C}$	2.0	−24	99	289	92	22	−2421(l)	9.80[a]
ethanamine (ethylamine)	$C_2H_5NH_2$	45.1	0.689$^{15\,°C}$(l)	−81.0	16.6	−16	1.363(l)	4.1	−47	36	284	72	27	−1713(l)	10.64[a]
N-ethylethanamine (diethylamine)	$(C_2H_5)_2NH$	73.1	0.699	−49.8	55.5	−28	1.383	3.1	−72	72	352	116	31	−3043(l)	10.93[a]
N,N-diethylthanamine (triethylamine)	$(C_2H_5)_3N$	101.2	0.723	−114.7	89.5	−6	1.398	2.2	−93	117	405	161	35	−4371(l)	10.72[a]
propan-1-amine	$CH_3CH_2CH_2NH_2$	59.1	0.712	−83.0	48.5	−44	1.385	3.9	−70	40	324	96	31	−2365(l)	10.57[a]
propan-2-amine	$CH_3CH(NH_2)CH_3$	59.1	0.686	−95.1	31.7	−50	1.371	4.0	−84	32	312	98	29	−2355(l)	10.63[a]
butan-1-amine	$CH_3(CH_2)_3NH_2$	73.1	0.733	−49.1	77.9	−12	1.399	3.3	−92	49	363	119	36	−3018(l)	10.64[a]
butan-2-amine	$CH_3CH_2CH(NH_2)CH_3$	73.1	0.720	−72	63	−19	1.391	4.3(b)	−104	41	351	117	33	−3009(l)	10.57[a]
2-methylpropan-1-amine (isobutylamine)	$CH_3CH(CH_3)CH_2NH_2$	73.1	0.728	−84.6	67.7	−20	1.395	4.2(b)	−99				34	−3014(l)	10.43[a]
2-methylpropan-2-amine (*tert*-butylamine)	$CH_3C(CH_3)(NH_2)CH_3$	73.1	0.691	−66.9	44.4	−8	1.376	4.3(b)	−121	28	338	120	31	−2996(l)	10.67[a]
aniline (aminobenzene)	$C_6H_5NH_2$	93.1	1.017	−6.0	184.1	70	1.584	5.1	87	167	319	108	56	−3393(l)	4.60[a]

[a] pK_a of the conjugate acid, BH^+.

Name	Formula	M / g mol^{-1}	ρ / g cm^{-3}	t_m / °C	t_b / °C	t_{FP} / °C	n	p / 10^{-30} C m (gas)	$\Delta_f H^\ominus$ / kJ mol^{-1} (gas)	$\Delta_f G^\ominus$ / kJ mol^{-1} (gas)	$S^\ominus$ / J K^{-1} mol^{-1} (gas)	$C_p^\ominus$ / J K^{-1} mol^{-1} (gas)	$\Delta_{vap}H$ / kJ mol^{-1}	$\Delta_c H^\ominus$ / kJ mol^{-1} (state at 298 K)	pK_a
Amines *(cont.)*															
2-chloroaniline	$ClC_6H_4NH_2$	127.6	1.213$^{20\,°C}$	–14	208.8	97	1.589$^{20\,°C}$	5.9(b)							0.79[a]
3-chloroaniline	$ClC_6H_4NH_2$	127.6	1.216$^{20\,°C}$	–10.3	229.9	123	1.594$^{20\,°C}$	8.9(b)							2.65[a]
4-chloroaniline	$ClC_6H_4NH_2$	127.6	1.429$^{19\,°C}$	72.5	232		1.555$^{87\,°C}$	9.9(b)							4.1[a]
2-methylaniline (*o*-toluidine)	$CH_3C_6H_4NH_2$	107.2	0.994	–16.1	200.4	85	1.570	5.3(b)	56	168	351	130	63	–4035(l)	4.45[a]
3-methylaniline (*m*-toluidine)	$CH_3C_6H_4NH_2$	107.2	0.985	–30.4	203.4	85	1.566	4.8(b)	55	165	353	126	61	–4033(l)	4.72[a]
4-methylaniline (*p*-toluidine)	$CH_3C_6H_4NH_2$	107.2	0.962$^{20\,°C}$	43.7	200.6	88	1.554$^{45\,°C}$	4.4(b)	55	168	347	126	79(sub)	–4017(s)	5.08[a]
2-nitroaniline	$NO_2C_6H_4NH_2$	138.1	1.442$^{15\,°C}$	71.5	284			14.1(b)	64				90(sub)	–3193(s)	–0.26[a]
3-nitroaniline	$NO_2C_6H_4NH_2$	138.1	1.430$^{4\,°C}$	114.0	d306			16(b)	58				97(sub)	–3180(s)	2.6[a]
4-nitroaniline	$NO_2C_6H_4NH_2$	138.1	1.424$^{20\,°C}$	147.5	332	165		19	59				101(sub)	–3177(s)	1.00[a]
N-phenylaniline (diphenylamine)	$C_6H_5NHC_6H_5$	169.2	1.160$^{20\,°C}$	52.9	302	152		3.6(b)	219	361	408		70	–6424(s)	0.79[a]
benzene-1,2-diamine (*o*-phenylenediamine)	$NH_2C_6H_4NH_2$	108.1		102.5	256.0			5.1	–0.3(s)					–3504(s)	0.6[a], 4.63[b]
benzene-1,3-diamine (*m*-phenylenediamine)	$NH_2C_6H_4NH_2$	108.1	1.010$^{58\,°C}$	63.5	282–4		1.634$^{58\,°C}$	6.0	–8(s)	169(s)	155(s)	160(s)		–3497(s)	2.5[a], 5.1[b]
benzene-1,4-diamine (*p*-phenylenediamine)	$NH_2C_6H_4NH_2$	108.1		146	267			5.1	3(s)					–3508(s)	3.0[a], 6.3[b]
Heterocyclic compounds															
pyrrole (azole)	C_4H_5N	67.1	0.970$^{20\,°C}$	–23.4	129.8	39	1.509$^{20\,°C}$	6.1	63(l)	149(l)	156(l)	128(l)	45	–2352(l)	0.4[a]
furan	C_4H_4O	68.1	0.951$^{20\,°C}$	–85.6	31.5	–35	1.421$^{20\,°C}$	2.2	–35	0.9	267	65	28	–2084(l)	
thiophene	C_4H_4S	84.1	1.057	–39.4	84.0	–1	1.526	1.8	115	126	279	73	35	–2828(l)	
pyridine (azine)	C_5H_5N	79.1	0.978	–42	115.5	20	1.507	7.3	140	190	283	78	40	–2782(l)	5.23[a]
2-methylpyridine (2-picoline)	$CH_3C_5H_4N$	93.1	0.940	–66.8	128.8	26	1.498	6.2	99	177	325	100	42	–3419(l)	5.95[a]

[a] pK_a of the conjugate acid, BH^+.
[b] pK_a of the acid, BH_2^{2+}.

Name	Formula	$\frac{M}{\text{g mol}^{-1}}$	$\frac{\rho}{\text{g cm}^{-3}}$	$\frac{t_m}{°C}$	$\frac{t_b}{°C}$	$\frac{t_{FP}}{°C}$	n	$\frac{p}{10^{-30}\text{C m}}$ (gas)	$\frac{\Delta_f H^{\ominus}}{\text{kJ mol}^{-1}}$ (gas)	$\frac{\Delta_f G^{\ominus}}{\text{kJ mol}^{-1}}$ (gas)	$\frac{S^{\ominus}}{\text{J K}^{-1}\text{mol}^{-1}}$ (gas)	$\frac{C_p^{\ominus}}{\text{J K}^{-1}\text{mol}^{-1}}$ (gas)	$\frac{\Delta_{vap}H}{\text{kJ mol}^{-1}}$	$\frac{\Delta_c H^{\ominus}}{\text{kJ mol}^{-1}}$ (state at 298 K)	pK_a
Heterocyclic compounds (*cont.*)						36									
3-methylpyridine (3-picoline)	$CH_3C_5H_4N$	93.1	0.952	−18.3	144.1		1.504	8.0(b)	106	184	325	100	44	−3427(l)	5.70[a]
4-methylpyridine (4-picoline)	$CH_3C_5H_4N$	93.1	0.950	3.6	145.3	56	1.503	9.0	102				45	−3418(l)	
indole	C_8H_7N	117.2	1.22	52.5	254	>110	$1.609^{60°C}$	6.8(b)	157				70(sub)	−4273(s)	−3.5[a]
1-benzofuran (coumarone)	C_8H_6O	118.1	1.091	<−18	174	56	$1.565^{23°C}$								
1-benzothiophene (thianaphthene)	C_8H_6S	134.2	$1.148^{32°C}$	32	221	>110	$1.637^{37°C}$	2.1(b)	101(s)	188(s)	177(s)		66(sub)		
adenine	$C_5H_5N_5$	135.1		d360	s220				96(s)	300(s)	151(s)	147(s)	109(sub)	−2279(s)	4.20[a], 9.87
cytosine	$C_4H_5N_3O$	111.1		d320					−221(s)					−2067(s)	4.6[a], 2.15
guanine	$C_5H_5N_5O$	151.1		d360	s				−184(s)	48(s)	160(s)			−2499(s)	3.3[a], 9.31
thymine	$C_5H_6N_2O_2$	126.1		316					−463(s)			151(s)	134(sub)	−2357(s)	9.90
uracil	$C_4H_4N_2O_2$	112.1		335					−429(s)			121(s)	127(sub)	−1716(s)	9.49
1H-imidazole	$C_3H_4N_2$	68.1	$1.03^{101°C}$	90.5	256									−1811(s)	
Carboxylic acids															
formic acid (methanoic acid)	HCOOH	46.0	1.214	8.4	100.8	43	1.369	4.7	−379	−351	249	45	46	−255(l)	3.74
acetic acid (ethanoic acid)	CH_3COOH	60.1	1.044	16.6	117.9	40	1.370	5.8	−433	−375	283	67	52	−875(l)	4.76
aminoacetic acid (glycine)	NH_2CH_2COOH	75.1	$1.161^{20°C}$	d292					−529(s)	−369(s)	104(s)	99(s)	136(sub)	−973(s)	9.78
bromoacetic acid	$BrCH_2COOH$	139.0	$1.934^{50°C}$	50.0	208.0	>110	$1.480^{50°C}$							−2.90	
chloroacetic acid	$ClCH_2COOH$	94.5	$1.404^{40°C}$	63.0	189.3		$1.435^{55°C}$		−435				75(sub)		2.86
dichloroacetic acid	$Cl_2CHCOOH$	128.9	$1.563^{20°C}$	13.5	194.0	>110	$1.466^{20°C}$		−501(l)						1.1
trichloroacetic acid	Cl_3CCOOH	163.4	1.62	57.5	196.5		$1.460^{61°C}$		−503(s)						−0.5

[a] pK_a of the conjugate acid, BH^+.

Name	Formula	M / g mol^{-1}	ρ / g cm^{-3}	t_m / °C	t_b / °C	t_{FP} / °C	n	p / 10^{-30} C m (gas)	$\Delta_f H^\circ$ / kJ mol^{-1} (gas)	$\Delta_f G^\circ$ / kJ mol^{-1} (gas)	S° / J K^{-1} mol^{-1} (gas)	C_p° / J K^{-1} mol^{-1} (gas)	$\Delta_{vap}H$ / kJ mol^{-1}	$\Delta_c H^\circ$ / kJ mol^{-1} (state at 298 K)	pK_a
Carboxylic acids (*cont.*)															
fluoroacetic acid	FCH_2COOH	78.0	1.369$^{36\,°C}$	35.2	168				−679(s)						2.59
hydroxyacetic acid (glycolic acid)	$OHCH_2COOH$	76.1		80	d				−662(s)					−695(s)	3.83
iodoacetic acid	ICH_2COOH	186.0		83	d										3.17
propanoic acid	CH_3CH_2COOH	74.1	0.988	−20.7	140.8	49	1.385	5.8	−454				57	−1527(l)	4.87
2-chloropropanoic acid	$CH_3CHClCOOH$	108.5	1.258$^{20\,°C}$		186	101	1.438$^{20\,°C}$		−525(l)						2.90
3-chloropropanoic acid	$ClCH_2CH_2COOH$	108.5		41	d204	>110			−550(s)						4.11
(2S)-2-hydroxypropanoic acid (S-lactic acid)	$CH_3CH(OH)COOH$	90.1	1.206	53		>110					143(s)				3.86
(2R)-2-hydroxypropanoic acid (R-lactic acid)	$CH_3CH(OH)COOH$	90.1		53		>110			−695(s)	−524(s)	142(s)			−1344(s)	3.86
2-hydroxypropanoic acid (lactic acid)	$CH_3CH(OH)COOH$	90.1	1.206	18		>110	1.439$^{20\,°C}$		−675(l)	−518(l)	192(l)				3.86
butanoic acid	$CH_3(CH_2)_2COOH$	88.1	0.953	−5.4	163.3	67	1.396		−534(l)	−376(l)	222(l)	179(l)	58	−2184(l)	4.82
2-chlorobutanoic acid	$CH_3CH_2CHClCOOH$	122.6	1.179$^{20\,°C}$				1.441$^{20\,°C}$		−577(l)						2.92
3-chlorobutanoic acid	$CH_3CHClCH_2COOH$	122.6	1.190$^{20\,°C}$	16		>110	1.422$^{20\,°C}$		−558(l)						4.17
4-chlorobutanoic acid	$ClCH_2(CH_2)_2COOH$	122.6	1.224$^{20\,°C}$	16		>110	1.464$^{20\,°C}$		−568(l)						4.52
2-methylpropanoic acid (isobutyric acid)	$CH_3CH(CH_3)COOH$	88.1	0.943	−46.0	154.7	55	1.391		−553(l)						4.85
pentanoic acid (valeric acid)	$CH_3(CH_2)_3COOH$	102.1	0.935	−34	185.5	88	1.406		−490	−357	440	210(l)	68	−2837(l)	4.84
2,2-dimethylpropaonic acid	$CH_3C(CH_3)_2COOH$	102.1	0.905$^{50\,°C}$	35	164	63	1.393		−491				73		5.03
hexanoic acid	$CH_3(CH_2)_4COOH$	116.2	0.923	−3	205.7	104	1.415		−512				72	−3492(l)	4.86
tetradecanoic acid (myristic acid)	$CH_3(CH_2)_{12}COOH$	228.4	0.862$^{54\,°C}$	53.9			1.427$^{70\,°C}$		−694			432(s)	95	−8677(s)	
hexadecanoic acid (palmitic acid)	$CH_3(CH_2)_{14}COOH$	256.4	0.853$^{62\,°C}$	63	351.5		1.433$^{60\,°C}$		−892(s)	−314(s)	452(s)	461(s)	101	−9978(s)	

Name	Formula	$\frac{M}{\text{g mol}^{-1}}$	$\frac{\rho}{\text{g cm}^{-3}}$	$\frac{t_m}{°C}$	$\frac{t_b}{°C}$	$\frac{t_{FP}}{°C}$	n	$\frac{p}{10^{-30}\text{C m}}$ (gas)	$\frac{\Delta_f H^{\ominus}}{\text{kJ mol}^{-1}}$ (gas)	$\frac{\Delta_f G^{\ominus}}{\text{kJ mol}^{-1}}$ (gas)	$\frac{S^{\ominus}}{\text{J K}^{-1}\text{ mol}^{-1}}$ (gas)	$\frac{C_p^{\ominus}}{\text{J K}^{-1}\text{ mol}^{-1}}$ (gas)	$\frac{\Delta_{vap}H}{\text{kJ mol}^{-1}}$	$\frac{\Delta_c H^{\ominus}}{\text{kJ mol}^{-1}}$ (state at 298 K)	pK_a
Carboxylic acids *(cont.)*															
octadecanoic acid (stearic acid)	$CH_3(CH_2)_{16}COOH$	284.5	$0.941^{20\,°C}$	68.8	d350.0		$1.430^{80\,°C}$		−948(s)			502(s)	104	−11280(s)	
icosanoic acid (arachic acid)	$CH_3(CH_2)_{18}COOH$	312.5	$0.824^{100\,°C}$	76.5	d328.0		$1.425^{100\,°C}$		−812				128	−12575(s)	
(9*Z*)-octadec-9-enoic acid (oleic acid)	$CH_3(CH_2)_7CH{=}CH(CH_2)_7COOH$	282.5	0.887	13.4			$1.458^{20\,°C}$		−783(l)					−11194(l)	
(9*Z*, 12*Z*)-octadeca-9, 12-dienoic acid (linoleic acid)	$C_6H_{12}{=}C_3H_4{=}C_8H_{15}COOH$	280.5	$0.902^{20\,°C}$	−5			$1.470^{20\,°C}$		−674(l)						4.77
(9*Z*, 12*Z*, 15*Z*)-octadeca-9, 12, 15-trienoic acid (linolenic acid)	$CH_3(CH_2CH{=}CH)_3(CH_2)_7COOH$	278.4	$0.916^{20\,°C}$	−11		>110	$1.480^{20\,°C}$								
(5*Z*, 8*Z*, 11*Z*, 14*Z*)-icosa-5, 8, 11, 14-tetraenoic acid (arachidonic acid)	$C_{19}H_{31}COOH$	304.5	$0.908^{20\,°C}$	−49.5			$1.482^{20\,°C}$								
benzoic acid	C_6H_5COOH	122.1	$1.266^{15\,°C}$	122.4	249.0	121	$1.504^{132\,°C}$	5.7(b)	−294	−214	369	103	91(sub)	−3227(s)	4.20
2-aminobenzoic acid (anthranilic acid)	$NH_2C_6H_4COOH$	137.1	$1.412^{20\,°C}$	*146–7	s				−296				105(sub)	−3354(s)	2.08[a]
3-aminobenzoic acid	$NH_2C_6H_4COOH$	137.1	$1.510^{4\,°C}$	173					−284				128(sub)	−3344(s)	3.1[a], 4.7
4-aminobenzoic acid	$NH_2C_6H_4COOH$	137.1	1.374	188–9					−297				116(sub)	−3342(s)	2.2[a], 4.7
2-chlorobenzoic acid	ClC_6H_4COOH	156.6	$1.544^{20\,°C}$	*142	s				−325				80(sub)		2.92
3-chlorobenzoic acid	ClC_6H_4COOH	156.6	1.496	*158	s				−342				82(sub)		3.83
4-chlorobenzoic acid	ClC_6H_4COOH	156.6		243					−341				88(sub)		3.99
2-hydroxybenzoic acid (salicylic acid)	OHC_6H_4COOH	138.1	$1.443^{20\,°C}$	159			1.565		−590(s)	−423(s)	178(s)		95(sub)	−3023(s)	2.97, 13.70
3-hydroxybenzoic acid	OHC_6H_4COOH	138.1	1.473	202–3					−585(s)	−417(s)	177(s)				4.08, 9.96

[a] pK_a of the conjugate acid, BH^+.

Name	Formula	M / g mol^{-1}	ρ / g cm^{-3}	t_m / °C	t_b / °C	t_{FP} / °C	n	p / 10^{-30} C m (gas)	$\Delta_f H^\ominus$ / kJ mol^{-1} (gas)	$\Delta_f G^\ominus$ / kJ mol^{-1} (gas)	$S^\ominus$ / J K^{-1} mol^{-1} (gas)	$C_p^\ominus$ / J K^{-1} mol^{-1} (gas)	$\Delta_{vap}H$ / kJ mol^{-1}	$\Delta_c H^\ominus$ / kJ mol^{-1} (state at 298 K)	pK_a
Carboxylic acids (*cont.*)															
4-hydroxybenzoic acid	OHC_6H_4COOH	138.1	1.46	214.5					−585(s)	−417(s)	176(s)				4.58, 9.46
2-methylbenzoic acid (*o*-toluic acid)	$CH_3C_6H_4COOH$	136.2	$1.062^{115\,°C}$	103.7	259		$1.512^{115\,°C}$		−417(s)			175(s)		−3875(s)	3.90
3-methylbenzoic acid (*m*-toluic acid)	$CH_3C_6H_4COOH$	136.2	$1.054^{112\,°C}$	111–3			1.509		−426(s)			164(s)		−3865(s)	4.27
4-methylbenzoic acid (*p*-toluic acid)	$CH_3C_6H_4COOH$	136.2		179.6	275				−429(s)			169(s)		−3862(s)	4.37
2-nitrobenzoic acid	$NO_2C_6H_4COOH$	167.1	$1.575^{20\,°C}$	147–8					−398(s)						2.18
3-nitrobenzoic acid	$NO_2C_6H_4COOH$	167.1	$1.494^{20\,°C}$	141–2					−414(s)						3.45
4-nitrobenzoic acid	$NO_2C_6H_4COOH$	167.1	$1.610^{20\,°C}$	*242	s				−427(s)						3.44
phenylacetic acid	$C_6H_5CH_2COOH$	136.2	$1.228^{6\,°C}$	76.7	265.5				−396(s)						4.31
Dicarboxylic acids															
ethanedioic acid (oxalic acid)	$(COOH)_2$	90.0	$1.900^{17\,°C}$		s157.0				−822(s)	−690(s)	110(s)	91(s)		−246(s)	1.25, 4.27
ethanedioc acid dihydrate (oxalic acid dihydrate)	$(COOH)_2{\cdot}2H_2O$	126.1	$1.653^{18\,°C}$	101.5					−1427(s)					−218(s)	
propanedioic acid (malonic acid)	$COOHCH_2COOH$	104.1	$1.619^{16\,°C}$	d135					−891(s)					−861(s)	2.85, 5.70
butanedioic acid (succinic acid)	$COOH(CH_2)_2COOH$	118.1	1.572	188	d235				−941(s)	−744(s)	167(s)	153(s)		−1491(s)	4.21, 5.64
(2*R*, 3*R*)-2, 3-dihydroxybutanedioic acid ((*R*, *R*)-tartaric acid)	COOHCH(OH)CH(OH)COOH	150.1	$1.760^{20\,°C}$	170					−1257(s)						3.04, 4.37
(2*S*, 3*S*)-2, 3-dihydroxybutanedioic acid ((*S*, *S*)-tartaric acid)	COOHCH(OH)CH(OH)COOH	150.1	$1.760^{20\,°C}$	170			1.496								2.98, 4.34
2, 3-dihydroxybutanedioic acid (tartaric acid)	COOHCH(OH)CH(OH)COOH	150.1	1.788	206.0					−1278(s)					−1141(s)	2.98, 4.34

Name	Formula	M / g mol^{-1}	ρ / g cm^{-3}	t_m / °C	t_b / °C	t_{FP} / °C	n	p / 10^{-30} C m (gas)	$\Delta_f H^\circ$ / kJ mol^{-1} (gas)	$\Delta_f G^\circ$ / kJ mol^{-1} (gas)	S° / J K^{-1} mol^{-1} (gas)	C_p° / J K^{-1} mol^{-1} (gas)	$\Delta_{vap}H$ / kJ mol^{-1}	$\Delta_c H^\circ$ / kJ mol^{-1} (state at 298 K)	pK_a
Dicarboxylic acids (*cont.*)															
(2*R*, 3*S*)-2,3-dihydroxybutanedioic acid ((*R*, *S*)-tartaric acid)	COOHCH(OH)CH(OH)COOH	150.1	$1.666^{20\,°C}$	165			1.5		−1260(s)						3.17, 4.91
pentanedioic acid (glutaric acid)	$COOH(CH_2)_3COOH$	132.1	1.42	97.8	d303		$1.419^{106\,°C}$		−960(s)					−2151(s)	4.33, 5.42
hexanedioic acid (adipic acid)	$COOH(CH_2)_4COOH$	146.1	1.360	153	337.5	196			−994(s)						4.42, 5.42
benzene-1,2-dicarboxylic acid (phthalic acid)	$COOHC_6H_4COOH$	166.1	$1.593^{20\,°C}$	*191	d				−782(s)	−592(s)	208(s)	188(s)			2.95, 5.41
(2*Z*)-but-2-enedioic acid (maleic acid)	COOHCH=CHCOOH	116.1	$1.590^{20\,°C}$	130.5					−789(s)	−630(s)	161(s)	137(s)			1.9, 6.23
(2*E*)-but-2-enedioic acid (fumaric acid)	COOHCH=CHCOOH	116.1	$1.635^{20\,°C}$	*300	s				−812(s)	−655(s)	168(s)	142(s)			3.05, 4.49
Acid derivatives															
formamide (methanamide)	$HCONH_2$	45.0	1.129	2.5	220	154	1.446	12.4	−194	−149	249	45	60	−2796(s)	
urea	NH_2CONH_2	60.1	$1.32^{18\,°C}$	132.7	d		1.484	15.3(b)	−246	−153	249	93(s)	88(sub)	−3224(s)	
acetamide (ethanamide)	CH_3CONH_2	59.1	$1.159^{20\,°C}$	81.3	221.2		1.428	12.5	−317(s)	−191(s)	115(s)	91(s)	79(sub)	−1355(s)	
N-phenylacetamide (acetanilide)	$CH_3CONHC_6H_5$	135.2	$1.219^{15\,°C}$	114.3	304.0	173		12.2(b)	−129				81(sub)	−1335(s)	
N-(2-nitrophenyl)acetamide (2-nitroacetanilide)	$CH_3CONHC_6H_4NO_2$	180.2	$1.419^{15\,°C}$	94.3				5.5(b)	−214(s)						
N-(4-nitrophenyl)acetamide (4-nitroacetanilide)	$CH_3CONHC_6H_4NO_2$	180.2		216.3					−238(s)						
acetic anhydride (ethanoic anhydride)	$(CH_3CO)_2O$	102.1	1.075	−73.1	140.0	54	1.388	9.3	−573	−474	390	99	52	−1807	
acetyl chloride (ethanoyl chloride)	CH_3COCl	78.5	$1.104^{21\,°C}$	−112	51–2	5	$1.389^{20\,°C}$	9	−244	−206	295	68	30		
acetonitrile	CH_3CN	41.1	0.777	−43.8	81.6	5	1.342	13.1	64	82	245	52	33	−1247	

Name	Formula	$\frac{M}{\text{g mol}^{-1}}$	$\frac{\rho}{\text{g cm}^{-3}}$	$\frac{t_m}{°C}$	$\frac{t_b}{°C}$	$\frac{t_{FP}}{°C}$	n	$\frac{p}{10^{-30}\text{C m}}$ (gas)	$\frac{\Delta_f H^\ominus}{\text{kJ mol}^{-1}}$ (gas)	$\frac{\Delta_f G^\ominus}{\text{kJ mol}^{-1}}$ (gas)	$\frac{S^\ominus}{\text{J K}^{-1}\text{ mol}^{-1}}$ (gas)	$\frac{C_p^\ominus}{\text{J K}^{-1}\text{ mol}^{-1}}$ (gas)	$\frac{\Delta_{vap} H}{\text{kJ mol}^{-1}}$	$\frac{\Delta_c H^\ominus}{\text{kJ mol}^{-1}}$ (state at 298 K)	pK_a
Acid derivatives *(cont.)*															
isocyanomethane (methyl isocyanide)	CH_3NC	41.1	$0.758^{4\,°C}$	−45	d59			12.8	149	166	247	53	32	−1333	
2-(acetyloxy)benzoic acid (aspirin)	$CH_3COOC_6H_4COOH$	180.2	1.35	135					−816(s)						
Esters															
methyl methanoate (methyl formate)	$HCOOCH_3$	60.1	$0.974^{20\,°C}$	−99.0	31.5	−32	1.342	5.9	−355	−303	301	67	31	−973	
ethyl methanoate (ethyl formate)	$HCOOC_2H_5$	74.1	0.916	−79.6	54.2	−19	1.358	6.4	−371				28		
methyl ethanoate (methyl acetate)	CH_3COOCH_3	74.1	0.927	−98	57	−10	1.359	5.7	−412	−325	324	86	34	−1592	
ethyl ethanoate (ethyl acetate)	$CH_3COOC_2H_5$	88.1	0.895	−83.6	77.1	−3	1.370	5.9	−444	−328	363	114	35	−2238	
propyl ethanoate (propyl acetate)	$CH_3COOC_3H_7$	102.1	0.883	−95	101.6	14	$1.384^{20\,°C}$	6.0(b)							
propan-2-yl ethanoate (isopropyl acetate)	$CH_3COOC_3H_7$	102.1	$0.872^{20\,°C}$	−73.4	88.6	11	$1.377^{20\,°C}$	6.2(b)	−482			199(l)	37	−2878	
butyl ethanoate (butyl acetate)	$CH_3COOC_4H_9$	116.2	0.876	−77.9	126.5	22	$1.394^{20\,°C}$	6.2(b)	−486			228(l)	44	−3547	
phenyl ethanoate (phenyl acetate)	$CH_3COOC_6H_5$	136.2	$1.078^{20\,°C}$		196	76	$1.503^{20\,°C}$	5.1(b)	−280				55		
methyl propanoate	$C_2H_5COOCH_3$	88.1	0.909	−87.5	79.8	−2	1.374	5.7(b)	−473(l)						
ethyl propanoate	$C_2H_5COOC_2H_5$	102.1	$0.892^{20\,°C}$	−73.9	99.1	12	$1.384^{20\,°C}$	5.8(b)	−464				39	−2894	
methyl butanoate	$C_3H_7COOCH_3$	102.1	$0.898^{20\,°C}$	−84.8	102.8	14	$1.388^{20\,°C}$	5.7(b)	−498(l)						
methyl benzoate	$C_6H_5COOCH_3$	136.2	1.084	−12.4	199.5	82	$1.516^{20\,°C}$	6.2(b)	−288			221(l)	56	−3948	
ethyl benzoate	$C_6H_5COOC_2H_5$	150.2	1.042	−34.7	212.4	84	1.504	6.7	370(l)						
methyl 2-hydroxybenzoate (methyl salicylate)	$HOC_6H_4COOCH_3$	152.2	1.181	−8	223	96	$1.535^{20\,°C}$	8.2(l)	−532(l)			249(l)			

Name	Formula	M / g mol^{-1}	ρ / g cm^{-3}	t_m / °C	t_b / °C	t_{FP} / °C	n	p / 10^{-30} C m (gas)	$\Delta_f H^\circ$ / kJ mol^{-1} (gas)	$\Delta_f G^\circ$ / kJ mol^{-1} (gas)	S° / J K^{-1} mol^{-1} (gas)	C_p° / J K^{-1} mol^{-1} (gas)	$\Delta_{vap}H$ / kJ mol^{-1}	$\Delta_c H^\circ$ / kJ mol^{-1} (state at 298 K)	pK_a
Aldehydes															
methanal (formaldehyde)	HCHO	30.0	$0.815^{-20\,°C}$	−92	−19.2			7.8	−109	−102	219	35	24	−571	
ethanal (acetaldehyde)	CH_3CHO	44.1	$0.783^{18\,°C}$	−123	20.1	−38	1.332(l)	9.0	−166	−133	264	55	26	−1192	
trichloroacetaldehyde (chloral)	CCl_3CHO	147.4	$1.512^{20\,°C}$	−57.5	97.8		$1.456^{20\,°C}$	6.5(b)	−197			151(l)	40		
trichloroacetaldehyde hydrate (chloral hydrate)	$CCl_3CH(OH)_2$	165.4	$1.908^{20\,°C}$	57	d98				−449			142(s)	128(sub)		
propanal	CH_3CH_2CHO	58.1	0.866	−80	47.9	−26	$1.364^{20\,°C}$	8.4	−186	−124	305	81	30	−1823	
butanal	$CH_3(CH_2)_2CHO$	72.1	0.796	−99	74.8	−6	1.377	9.1	−205	−115	344	103	34	−2478	
3-hydroxybutanal (aldol)	$CH_3CH(OH)CH_2CHO$	88.1	$1.103^{20\,°C}$	d85			$1.424^{20\,°C}$		−431(l)						
2-methylpropanal	$CH_3CH(CH_3)CHO$	72.1	0.784	−65	64.1	−40	1.370		−216				32	−2470	
pentanal	$CH_3(CH_2)_3CHO$	86.1	0.805	−91	103	12	1.392		−228	−109	383	125	39	−3129	
hexanal	$CH_3(CH_2)_4CHO$	100.2	0.810	−56	131	30	1.402		−248	−100	423	148			
benzaldehyde	C_6H_5CHO	106.1	$1.045^{20\,°C}$	−26	178.9	62	$1.546^{20\,°C}$	9.9(l)	−37	22	336	172(l)	50	−3525	
prop-2-enal (acrylaldehyde)	$CH_2{=}CHCHO$	56.1	$0.841^{20\,°C}$	−87.7	52.6	−18	$1.402^{20\,°C}$	10.4	−122(l)						
(2*E*)-but-2-enal (*trans*-crotonaldehyde)	$CH_3CH{=}CHCHO$	70.1	0.849	−76	104–5	8	$1.437^{20\,°C}$	12.2	−101				38	−2293	
ethanedial (glyoxal)	CHOCHO	58.0	$1.14^{20\,°C}$	15	50.4		$1.383^{20\,°C}$	16.0	−212					−861(g)	
Ketones															
propanone (acetone)	CH_3COCH_3	58.1	0.785	−94.8	56.1	−20	1.356	9.6	−218	−153	295	75	31	−1790	
butan-2-one (ethyl methyl ketone)	$CH_3CH_2COCH_3$	72.1	0.800	−86.6	79.6	−1	1.376	9.3	−238	−146	338	103	35	−2444	
pentan-2-one	$CH_3(CH_2)_2COCH_3$	86.1	0.802	−77.8	102.3	7	1.389	9.2(b)	−259	−137	376	121	38	−3099	

Name	Formula	M / g mol^{-1}	ρ / g cm^{-3}	t_m / °C	t_b / °C	t_{FP} / °C	n	p / 10^{-30}C m (gas)	$\Delta_f H^\ominus$ / kJ mol^{-1} (gas)	$\Delta_f G^\ominus$ / kJ mol^{-1} (gas)	$S^\ominus$ / J K^{-1} mol^{-1} (gas)	$C_p^\ominus$ / J K^{-1} mol^{-1} (gas)	$\Delta_{vap}H$ / kJ mol^{-1}	$\Delta_c H^\ominus$ / kJ mol^{-1} (state at 298 K)	pK_a
Ketones (*cont.*)															
3-methylbutan-2-one	$CH_3CH(CH_3)COCH_3$	86.1	0.798	−92.0	94.3	6	1.386	9.2(b)	−299(l)	−146(l)	268(l)	180(l)	37	−3097	
pentan-3-one (diethyl ketone)	$CH_3CH_2COCH_2CH_3$	86.1	0.809	−39.5	101.8	7	1.390	9.1(b)	−258	−134	370	191(l)	39	−3100	
hexan-2-one	$CH_3(CH_2)_3COCH_3$	100.2	0.807	−55.8	127.2	23	1.399	8.9(b)	−280			213(l)	42	−3754	
4-methylpentan-2-one (methyl isobutyl ketone)	$CH_3CH(CH_3)CH_2COCH_3$	100.2	0.796	−84	116.2	13	1.394		−389(l)						
3,3-dimethylbutan-2-one (pinacolone)	$(CH_3)_3CCOCH_3$	100.2	0.723	−52.5	106	23	1.394	9.0(b)	−291				38	−3747	
cyclobutanone	$CH_2(CH_2)_2CO$ (cyclic)	70.1	0.928	−50.9	99	16	1.419	9.6							
cyclopentanone	$CH_2(CH_2)_3CO$ (cyclic)	84.1	0.944	−51.3	130.6	30	1.435	11.0	−192				44	−2875	
cyclohexanone	$CH_2(CH_2)_4CO$ (cyclic)	98.1	0.942	−31	155.7	46	$1.451^{20\,°C}$	9.6	−226	−87	322	110	45	−3579	
butane-2,3-dione (biacetyl)	$CH_3COCOCH_3$	86.1	$0.981^{18\,°C}$	−2.4	88	26	$1.395^{20\,°C}$	4.2	−327				39	−2066	
pentane-2,4-dione (acetylacetone)	$CH_3COCH_2COCH_3$	100.1	0.972	−23.0	141	30	1.449	10.1	−381				43	−2687	
1-phenylethanone (acetophenone)	$C_6H_5COCH_3$	120.1	1.024	19.7	202	82	1.532	10.1	−87	2	373		56	−4149	
diphenylmethanone (benzophenone)	$C_6H_5COC_6H_5$	182.2	$1.146^{20\,°C}$	48.1	305.9	>110	$1.608^{19\,°C}$	8.3	55			225(s)	89(sub)	−6510	
ethenone (ketene)	$CH_2{=}CO$	42.0		−151	−50			4.7	−48	−48	248	52	20	−1025	
Ethers															
methoxymethane (dimethyl ether)	CH_3OCH_3	46.1		−141.5	−24.9	−83	1.302(l)	4.3	−184	−113	267	64	22	−1460	
methoxyethane (ethyl methyl ether)	$CH_3OC_2H_5$	60.1	$0.725^{0\,°C}$	−113	7.4		1.342(l)	4.1	−216	−118	311	90		−2108	
ethoxyethane (diethyl ether)	$C_2H_5OC_2H_5$	74.1	0.708	−116.3	34.5	−40	1.350	3.8	−252	−122	343	113	27	−2724	

Name	Formula	M / g mol^{-1}	ρ / g cm^{-3}	t_m / °C	t_b / °C	t_{FP} / °C	n	p / 10^{-30}C m (gas)	$\Delta_f H^\ominus$ / kJ mol^{-1} (gas)	$\Delta_f G^\ominus$ / kJ mol^{-1} (gas)	$S^\ominus$ / J K^{-1} mol^{-1} (gas)	$C_p^\ominus$ / J K^{-1} mol^{-1} (gas)	$\Delta_{vap}H$ / kJ mol^{-1}	$\Delta_c H^\ominus$ / kJ mol^{-1} (state at 298 K)	pK_a
Ethers *(cont.)*															
1-methoxypropane	$CH_3OC_3H_7$	74.1	$0.738^{20\,°C}$		39.1		1.358	3.7	−238	−110	349	113	28	−2737	
2-methoxypropane	$CH_3OCH(CH_3)CH_3$	74.1	$0.724^{15\,°C}$		30.7		$1.358^{20\,°C}$		−252	−121	338	111	27	−2724	
1-methoxybutane	$CH_3OC_4H_9$	88.2	$0.744^{20\,°C}$	−115.5	71		$1.374^{20\,°C}$	4.2(b)	−291(l)	−106(l)	295(l)	193(l)	32	−3392	
2-methoxy-2-methylpropane	$CH_3C(CH_3)(CH_3)OCH_3$	88.2	$0.740^{20\,°C}$	−109	55	−28	$1.369^{20\,°C}$	4.5(b)	−293	−125	353	134	30	−3360	
methoxybenzene (anisole)	$CH_3OC_6H_5$	108.1	0.989	−37.5	153.8	51	1.514	4.6	−68				47	−3783	
ethoxybenzene	$C_2H_5OC_6H_5$	122.2	0.960	−29.5	170	57	1.505	4.8	−102			229(l)	51	−4425	
diphenyl ether	$C_6H_5OC_6H_5$	170.2	$1.066^{30\,°C}$	26.8	258.3	115	1.579	4.3	−32(s)	144(s)	234(s)	217(s)	84(sub)	−6119(s)	
oxirane (ethylene oxide)	CH_2OCH_2	44.1	$0.891^{5\,°C}$	−111.7	10.6	−30	$1.358^{7\,°C}$	6.3	−53	−13	242	48	25	−1306	
tetrahydrofuran	$CH_2CH_2OCH_2CH_2$	72.1	$0.889^{20\,°C}$	−108	66	−17	1.405	5.4	−184	−81	302	76	32	−2501	
1,4-dioxane	$CH_2OCH_2CH_2OCH_2$	88.1	1.028	11.8	101.4	12	1.420	0	−316	−181	300	94	38	−2364	
2-methoxyethanol	$CH_3OCH_2CH_2OH$	76.1	$0.965^{20\,°C}$	−85.1	124.1	46	$1.402^{20\,°C}$		−480(l)						
Carbohydrates															
glucose, D(+)-	$CHOH(CHOH)_3CHOCH_2OH$	180.2	$1.562^{18\,°C}$	d146					−1273(s)	−911(s)	212(s)	219(s)		−2803(s)	
fructose, D(−)-	$OC(OH)(CH_2OH)(CHOH)_2CHCH_2OH$	180.2	$1.60^{20\,°C}$	d103					−1266(s)	−915(s)		230(s)		−2810(s)	
galactose, D(+)-	$O(CHOH)_4CHCH_2OH$	180.2		167					−1286(s)	−920(s)	205(s)	216(s)		−2790(s)	
mannose, D(+)-	$O(CHOH)_4CHCH_2OH$	180.2	$1.539^{20\,°C}$	d132					−1263(s)			214(s)		−2813(s)	
sucrose, D(+)- (glucose + fructose)	$(C_6H_{11}O_5)O(C_6H_{11}O_5)$	342.3	1.587	186			1.538		−2226(s)	−1548(s)	360(s)	425(s)		−5640(s)	
maltose, D(+)- (glucose + glucose)	$(C_6H_{11}O_5)O(C_6H_{11}O_5)$	342.3		160					−2221(s)	−1726(s)		435(s)		−5646(s)	
lactose, D(+)- (galactose + glucose)	$(C_6H_{11}O_5)O(C_6H_{11}O_5)$	342.3	$1.59^{20\,°C}$	253					−2237(s)	−1567(s)	386(s)	418(s)		−5630(s)	

26 PROPERTIES OF AMINO ACIDS

The following table contains symbols, names, formulae, solubility, thermochemical data, acid dissociation constants and the pH at the isoelectric points of common amino acids.

Name International Union of Pure and Applied Chemistry systematic nomenclature is used.

Formula The amino acids are depicted using condensed structural formulae.

M (in g mol^{-1}) = molar mass for the formula shown.

ρ (in g cm^{-3}) = density of the solid or liquid form at 298 K and 101.325 kPa, unless otherwise specified. Reported densities followed by (l) refer to the liquid form at a pressure greater than 101.325 kPa at 298 K. Reported densities marked * refer to the liquid at a pressure greater than 101.325 kPa at the specified temperature.

t_m (in °C) = melting temperature. (d = decomposes; s = sublimes; * = melts under pressure; values in parentheses are estimated.)

sol. (in g/100 g) = solubility. This is reported as the mass in g of anhydrous substance per 100 g of water at 298 K.

$\Delta_f H^\ominus$ (in kJ mol^{-1}) = the standard enthalpy of formation for 1 mole of the substance, in the gas phase unless otherwise indicated, under standard state conditions, from its elements in their standard reference states (stable forms).

$\Delta_f G^\ominus$ (in kJ mol^{-1}) = the standard Gibbs energy of formation for 1 mole of the substance, in the gas phase unless otherwise indicated, under standard state conditions, from its elements in their standard reference states (stable forms).

$S^\ominus$ (in J K^{-1} mol^{-1}) = the standard entropy of the substance in the gas phase unless otherwise indicated.

$C_p^\ominus$ (in J K^{-1} mol^{-1}) = the standard molar heat capacity at constant pressure of the substance in the gas phase unless otherwise indicated.

$\Delta_{sub}H$ (in kJ mol^{-1}) = the molar enthalpy of sublimation at 298 K for the transition solid to vapour.

$\Delta_c H^\ominus$ (in kJ mol^{-1}) = the standard molar enthalpy of combustion, for the substance in its stable form at 10^5 Pa and 298 K unless otherwise indicated, the final products being CO_2(g), H_2O(l), N_2(g) and H_2SO_4(aq, 1 mol H_2SO_4 in 115 mol H_2O). In calculating an enthalpy of formation from an enthalpy of combustion, or vice versa, the following standard enthalpies of formation should be used: CO_2(g), −393.5 kJ mol^{-1}; H_2O(l), −285.8 kJ mol^{-1}; H_2SO_4(aq, in 115 mol H_2O), −887.8 kJ mol^{-1}.

pK_a = −log K_a; K_a = the dissociation constant of the acid at 298 K.

p$K_{a,1}$ = −log of the first K_a.

p$K_{a,2}$ = −log of the second K_a.

pI = the pH at the isoelectric point. The pI is the pH at which the amino acid has no net charge. For example, glycine can exist in the following three different protonation states.

$$H_3\overset{+}{N}CH_2COOH \underset{}{\overset{pK_{a,1}}{\rightleftharpoons}} H_3\overset{+}{N}CH_2COO^- \overset{pK_{a,2}}{\rightleftharpoons} H_2NCH_2COO^-$$

p$K_{a,1}$ = 2.35 and p$K_{a,2}$ = 9.78, hence

$$\begin{aligned} pI &= \tfrac{1}{2}(pK_{a,1} + pK_{a,2}) \\ &= \tfrac{1}{2}(2.35 + 9.78) \\ &= 6.06 \end{aligned}$$

Symbol/Name/Code/Formula				M g mol^{-1}	ρ g cm^{-3}	t_m °C	*sol.* g/100 g	$\Delta_f H^{\circ}$ kJ mol^{-1} (solid)	$\Delta_f G^{\circ}$ kJ mol^{-1} (solid)	S° J K^{-1} mol^{-1} (solid)	C_p° J K^{-1} mol^{-1} (solid)	$\Delta_{sub}H$ kJ mol^{-1} (solid)	$\Delta_c H^{\circ}$ kJ mol^{-1} (solid)	p$K_{a,1}$ α–COOH	p$K_{a,2}$ α–NH_3^+	p$K_{a,x}$ side chain	pI
Ala	alanine	A	$C_3H_7NO_2$	89.1	1.424	d297	16.7	−561	−369	129	122	138	−1620	2.34	9.87		6.10
Arg	arginine	R	$C_6H_{14}N_4O_2$	174.2	1.100	d238	18.1	−624	−240	251	232		−3738	1.82	8.99	12.48	10.74
Asn	asparagine	N	$C_4H_8N_2O_3$	132.1	1.54$^{15\,°C}$	d236	2.5	−789	−530	174	160		−1928	2.1	8.7		5.4
Asp	aspartic acid	D	$C_4H_7NO_4$	133.1	1.660$^{13\,°C}$	d270	0.5	−973	−730	170	155		−1601	1.99	10.00	3.90	2.95
Cys	cysteine	C	$C_3H_7NO_2S$	121.2		d240	vs	−516	−326	170	162		−2267	1.92	8.35	10.46	5.13
Gln	glutamine	Q	$C_5H_{10}N_2O_3$	146.2		d185	4.2	−826	−532	195	184		−2570	2.17	9.13		5.65
Glu	glutamic acid	E	$C_5H_9NO_4$	147.1	1.538$^{20\,°C}$	d249	0.86	−1005	−727	188	175		−2248	2.16	9.93	4.27	3.22
Gly	glycine	G	$C_2H_5NO_2$	75.1	1.161$^{20\,°C}$	d292	25.1	−529	−369	104	99	136	−973	2.35	9.78		6.06
His	histidine	H	$C_6H_9N_3O_2$	155.2		d277	4.3	−467					−3181	1.82	9.17	6.00	7.59
Ile	isoleucine	I	$C_6H_{13}NO_2$	131.2		d284	3.4	−638	−347	208	188		−3581	2.32	9.76		6.04

All naturally occurring, chiral amino acids, with the exception of cysteine, display an *S* configuration at the α-carbon atom.

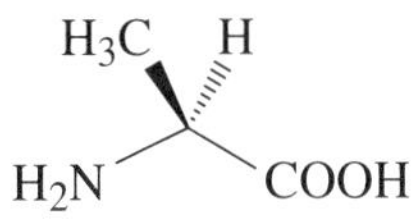

alanine (Ala, A)

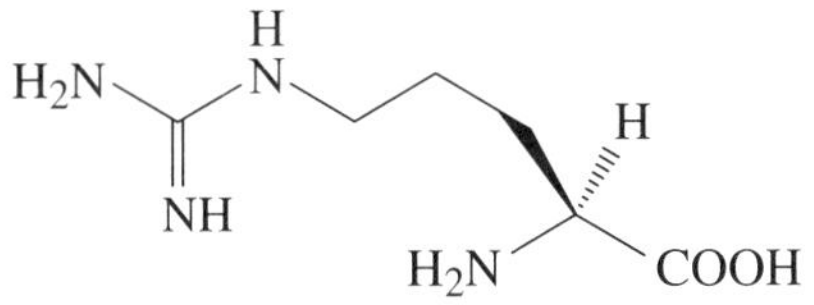

arginine (Arg, R)

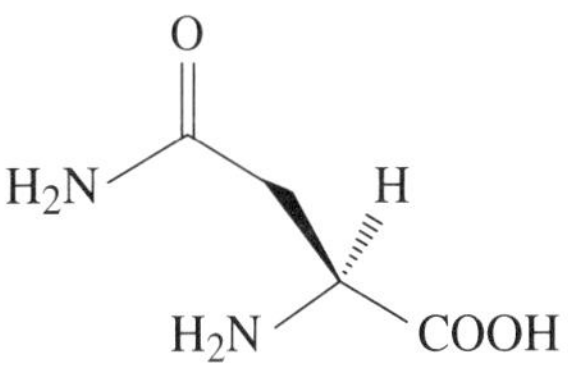

asparagine (Asn, N)

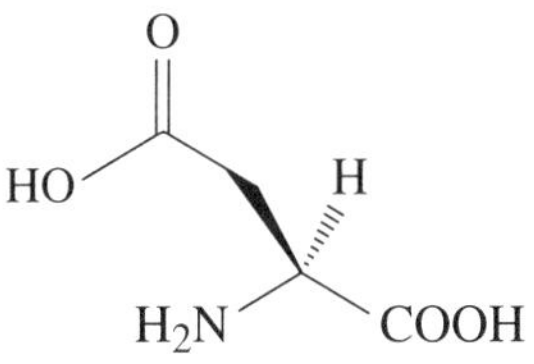

aspartic acid (Asp, D)

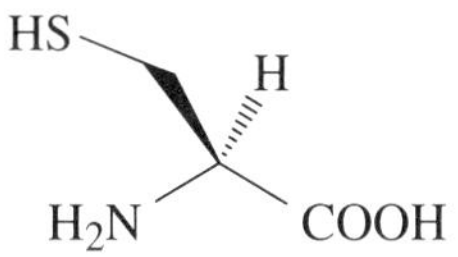

cysteine (Cys, C)

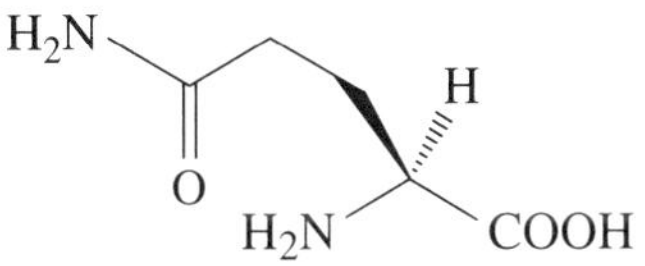

glutamine (Gln, Q)

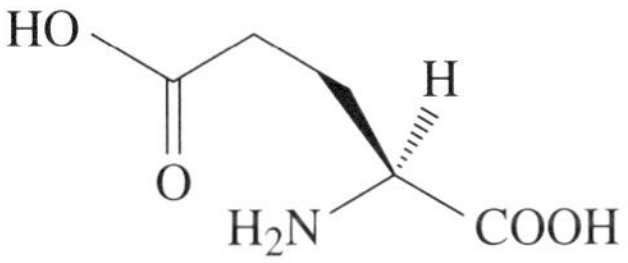

glutamic acid (Glu, E)

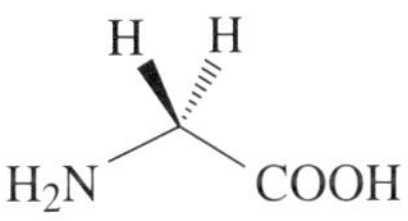

glycine (Gly, G)

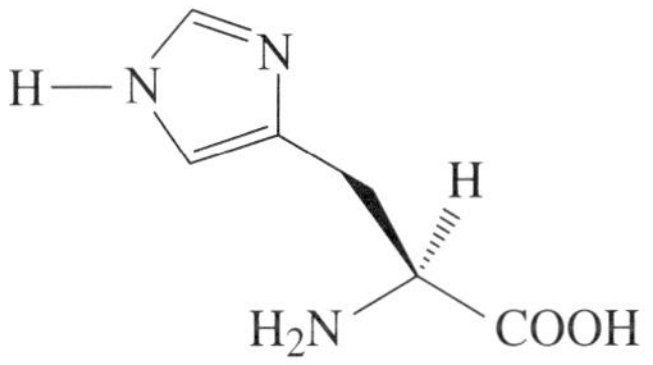

histidine (His, H)

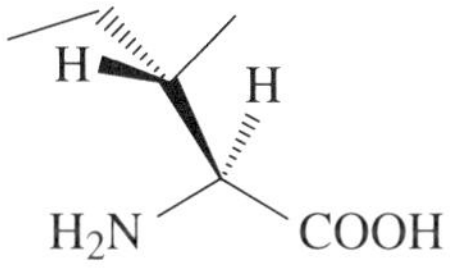

isoleucine (Ile, I)

26 PROPERTIES OF AMINO ACIDS *(continued)*

Symbol/Name/Code/Formula				M g mol^{-1}	ρ g cm^{-3}	t_m °C	*sol.* g/100 g	$\Delta_f H^\ominus$ kJ mol^{-1} (solid)	$\Delta_f G^\ominus$ kJ mol^{-1} (solid)	S° J K^{-1} mol^{-1} (solid)	C_p° J K^{-1} mol^{-1} (solid)	$\Delta_{sub}H$ kJ mol^{-1} (solid)	$\Delta_c H^\ominus$ kJ mol^{-1} (solid)	p$K_{a,1}$ α–COOH	p$K_{a,2}$ α–NH_3^+	p$K_{a,x}$ side chain	pI
Leu	leucine	L	$C_6H_{13}NO_2$	131.2	1.29$^{18\,°C}$	d337	2.3	−637	−347	212	201	151	−3581	2.33	9.74		6.04
Lys	lysine	K	$C_6H_{14}N_2O_2$	146.2		d224	0.6	−679					−3683	2.18	8.95	10.53	9.74
Met	methionine	M	$C_5H_{11}NO_2S$	149.2	1.340	d283	3.5	−759	−506	231	290		−3669	2.28	9.21		5.74
Phe	phenylalanine	F	$C_9H_{11}NO_2$	165.2		d284	2.9	−467	−211	214	203	154	−4647	2.16	9.18		5.67
Pro	proline	P	$C_5H_9NO_2$	115.1		d222	162	−512	−287	164	151	146	−2742	1.95	10.64		6.30
Ser	serine	S	$C_3H_7NO_3$	105.1	1.603$^{22\,°C}$	d228	25$^{20\,°C}$	−733	−515	149	136		−1448	2.19	9.21		5.70
Thr	threonine	T	$C_4H_9NO_3$	119.1		d253	20.5	−807	−550	153	142		−2053	2.09	9.10		5.60
Trp	tryptophan	W	$C_{11}H_{12}N_2O_2$	204.2		d282	1.1	−415	−119	251	238		−5628	2.43	9.44		5.94
Tyr	tyrosine	Y	$C_9H_{11}NO_3$	181.2		d344	0.05	−685	−399	214	216		−4429	2.20	9.11	10.13	5.66
Val	valine	V	$C_5H_{11}NO_2$	117.2	1.230	d315	8.9	−618	−358	179	169	163	−2922	2.32	9.62		5.97

All naturally occurring, chiral amino acids, with the exception of cysteine, display an *S* configuration at the α-carbon atom.

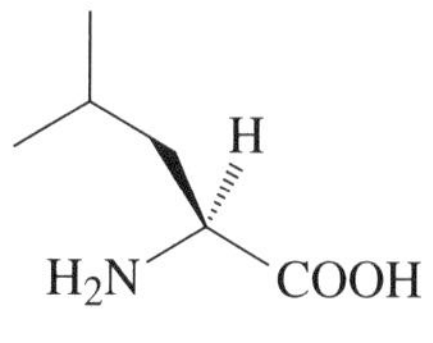

leucine (Leu, L)

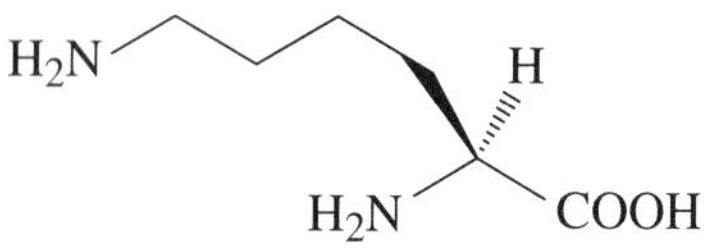

lysine (Lys, K)

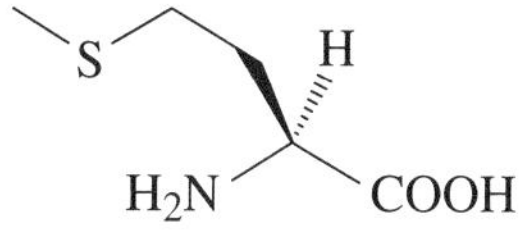

methionine (Met, M)

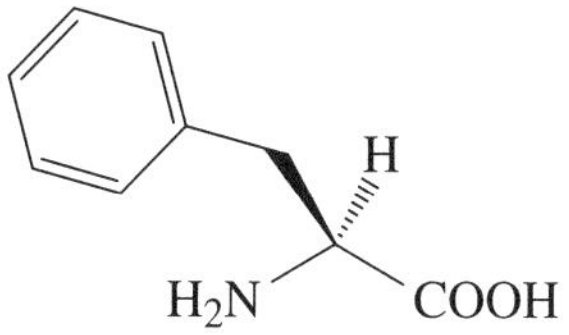

phenylalanine (Phe, F)

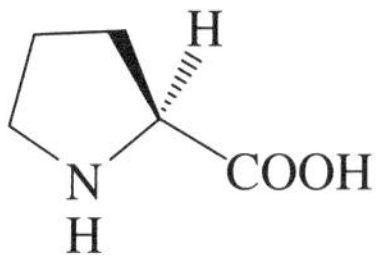

proline (Pro, P)

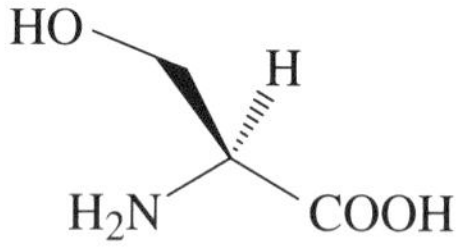

serine (Ser, S)

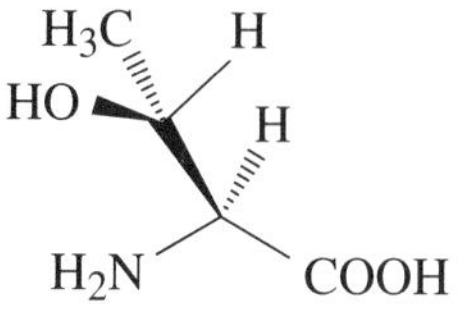

threonine (Thr, T)

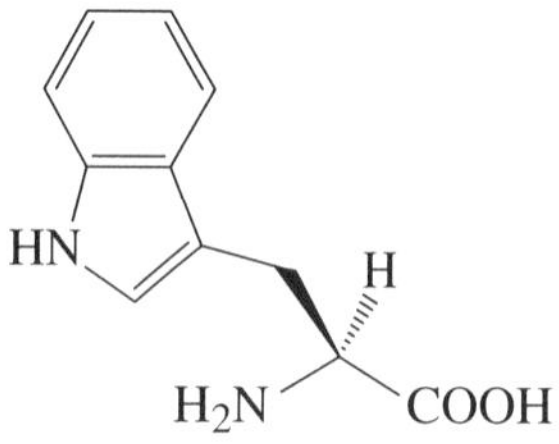

tryptophan (Trp, W)

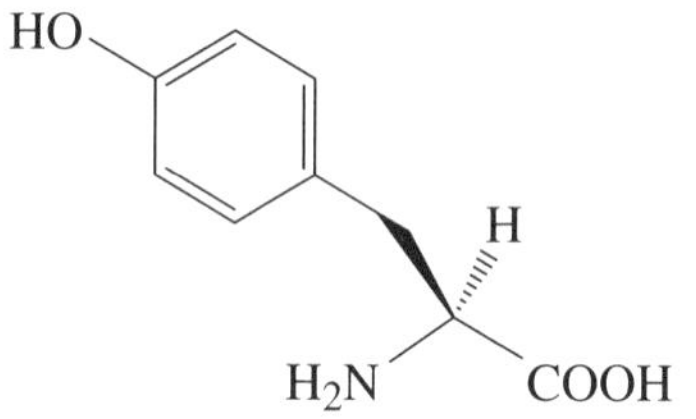

tyrosine (Tyr, Y)

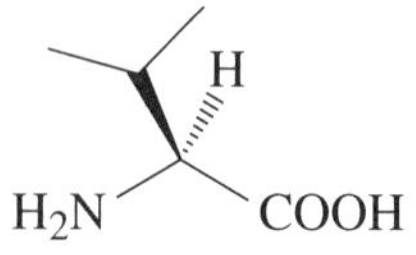

valine (Val, V)

27 PROPERTIES OF SOLVENTS

Solvent	Polarity index[a]	Viscosity (mPa s) at 20 °C	Boiling point (°C) at 101.325 kPa	Refractive index	UV cut off (nm)[b]
cyclohexane	0	0.98	80.7	1.427	200
hcxanc	0	0.31	68.7	1.372	195
decane	0.3	0.92	174.1	1.412	210
octane	0.4	0.50	125.7	1.404	210
diisopropyl ether	2.2	0.33	68.3	1.368	220
toluene	2.3	0.59	110.6	1.496	285
p-xylene	2.4	0.7	138.4	~1.50	290
benzene	3.0	0.65	80.1	1.501	280
dichloromethane	3.4	0.44	39.8	1.424	233
chloroform	3.4	0.57	61.1	1.443	245
1,2-dichloroethane	3.7	0.79	83.5	1.445	230
tetrahydrofuran	4.2	0.55	66.0	1.408	230
ethyl acetate	4.3	0.47	77.1	1.37	256
propan-1-ol	4.3	2.3	97.2	1.38	210
propan-2-ol	4.3	2.35	82.3	1.38	205
methyl acetate	4.4	0.37–0.45	57	1.362	260
ethylmethyl ketone	4.5	0.43	79.6	1.381	330
p-dioxane	4.8	1.54	101.4	1.422	215
ethanol	5.2	1.2	78.3	1.361	210
pyridine	5.3	0.94	115.5	1.51	330
acetone	5.4	0.32	56.1	1.359	330
2-methoxyethanol	5.7	1.72	124.1	1.401	220

[a] The polarity index is a relative measure of the degree of interaction of the solvent with various and different polar test solutes. It is a measure of the relative polarity of a solvent, and increases with polarity. It is useful for identifying suitable mobile phase solvents in chromatography.

[b] The UV cutoff is defined as the wavelength at which the absorbance of the given solvent in a 1 cm pathlength cell is equal to 1, when water is in a matched cell in the reference beam.

27 PROPERTIES OF SOLVENTS *(continued)*

Solvent	Polarity index[a]	Viscosity (mPa s) at 20 °C	Boiling point (°C) at 101.325 kPa	Refractive index	UV cut off (nm)[b]
acetic acid	6.2	1.26	117.9	1.372	230
acetonitrile	6.2	0.37	81.6	1.344	190
dimethylformamide	6.4	0.92	153	1.431	268
dimethylsulfoxide	6.5	2.24	189	1.477	268
methanol	6.6	0.6	64.7	1.329	205
formamide	7.3	3.76	220.0	1.446	~260
water	9.0	1.0	100	1.33	190

28 MISCIBILITY OF SOLVENTS

□ = miscible ☒ = immiscible

	acetic acid	acetone	acetonitrile	benzene	butan-1-ol	carbon tetrachloride	chloroform	cyclohexane	cyclopentane	dichloromethane	dimethylformamide	dimethyl sulfoxide	dioxane	ethyl acetate	ethanol	diethyl ether	heptane	hexane	methanol	methyl ethyl ketone	pentane	isopropanol	tetrahydrofuran	toluene	water
acetic acid																									
acetone																									
acetonitrile								X	X								X	X			X				
benzene																									X
butan-1-ol																									X
carbon tetrachloride																									X
chloroform																									X
cyclohexane			X								X	X							X						X
cyclopentane			X								X	X							X						X
dichloromethane																									X
dimethylformamide								X	X								X	X			X				
dimethyl sulfoxide								X	X							X	X	X			X				
dioxane																									
ethyl acetate																									X
ethanol																									
diethyl ether												X													X
heptane			X								X	X							X						X
hexane			X								X	X							X						X
methanol								X	X								X	X							
methyl ethyl ketone																									X
pentane			X					X	X										X						X
isopropanol																									
tetrahydrofuran																									
toluene																									X
water				X	X	X	X	X	X	X				X		X	X	X		X	X			X	

29 BOILING TEMPERATURE ELEVATION AND FREEZING TEMPERATURE DEPRESSION CONSTANTS (EBULLIOSCOPIC AND CRYOSCOPIC CONSTANTS)

The change in temperature is given by the equations $\Delta T = K_b b$ and $\Delta T = K_f b$, where values of K_b and K_f are given below and b is the molality of solute particles in the given solvent. (Note: t_b is at 101.325 kPa.)

Solvent	$\frac{t_b}{°C}$	$\frac{K_b}{K\ kg\ mol^{-1}}$	$\frac{t_f}{°C}$	$\frac{K_f}{K\ kg\ mol^{-1}}$
Acetone	56.1	1.71	−94.8	2.40
Benzene	80.1	2.53	5.5	5.085
Camphor	207.4	5.611	178.8	40
Carbon tetrachloride	76.7	5.03	−23	29.8
Chloroform	61.1	3.63	−63.5	4.90
Cyclohexane	80.7	2.79	6.6	20.0
Ethanol	78.3	1.22	−114.1	
Methanol	64.7	0.83	−97.7	
Naphthalene	217.9	5.80	80.3	6.94
Water	100.00	0.512	0.0	1.853

30 CRITICAL CONSTANTS AND TRIPLE POINTS

The triple point is the temperature and pressure where three phases are in equilibrium. In this table, data are given for the solid–liquid–gas triple point.

t_c (in °C) = critical temperature (This is the temperature above which the substance cannot exist in the liquid state.)
P_c (in MPa) = critical pressure
ρ_c (in g cm^{-3}) = critical density
t_t (in °C) = triple point temperature
P_t (in kPa) = triple point pressure

Substance	t_c / °C	P_c / MPa	ρ_c / g cm^{-3}	t_t / °C	P_t / kPa
air	−141	3.77	0.316	−214.0	8.1
Ar	−122	4.90	0.531	−189.3	68.8
Br_2	315	10.3	1.24	−7.3	6.09
C (graphite)	6537	223	0.638	3974	1.0×10^4
CCl_2F_2 (CFC-12)	112	4.14	0.558	−158.0	9.3×10^{-5}
CCl_3F (CFC-11)	198	4.41	0.554	−111.1	4.7×10^{-3}
CCl_4	283	4.52	0.558	−22.6	1.05
CH_4	−83	4.60	0.163	−182.5	11.7
CO_2	31	7.38	0.468	−56.6	518
C_2H_2	35	6.14	0.231	−80.8	126
CF_3CH_2F (HFC-134a)	101	4.07	0.515		
C_2H_4	9	5.04	0.218	−169.2	0.14
C_2H_5OH	241	6.13	0.276	−114.1	7.2×10^{-7}
C_2H_6	32	4.88	0.203	−182.8	1.1×10^{-3}
$(C_2H_5)_2O$	194	3.64	0.265	−116.3	4.1×10^{-4}
C_6H_6	289	4.90	0.302	5.5	4.81
Cl_2	144	8.00	0.573	−103.1	1.19
F_2	−129	5.22	0.58	−219.6	0.25
HBr	90	8.55	0.81	−88.0	27.4

30 CRITICAL CONSTANTS AND TRIPLE POINTS *(continued)*

Substance	t_c / °C	P_c / MPa	ρ_c / $g\,cm^{-3}$	t_t / °C	P_t / kPa
HCl	51	8.31	0.45	−114.1	13.5
HF	188	6.48	0.29	−83.4	0.34
HI	151	8.2	1.05	−50.8	49.3
H_2	−240	1.29	0.031	−259.2	7.20
H_2O	374	22.1	0.325	0.007	0.611
H_2S	100	8.94	0.34	−85.6	22.7
He[a]	−268	0.23	0.070	−271.4[a]	1.5[a]
I_2	546	11.7	1.64	114.1	12.0
NH_3	132	11.4	0.235	−77.7	6.12
NO	−93	6.48	0.52	−163.7	21.9
N_2	−147	3.39	0.313	−210.0	12.5
N_2O	36	7.26	0.453	−90.9	87.8
N_2O_4	158	10.1	0.557	−11.3	15.4
Ne	−229	2.76	0.484	−248.6	43.2
O_2	−119	5.04	0.436	−218.8	0.15
SO_2	158	7.88	0.524	−75.5	0.17

[a] The triple point for helium is for two liquids plus gas.

31 THE SPECTRAL ENERGIES

Radiation	Wavelength (λ)/nm	Energy/cm^{-1}	Energy/kJ mol^{-1}
Radiowave	3×10^{12} to 3×10^{8}	3×10^{-6} to 3×10^{-2}	3.98×10^{-8} to 3.98×10^{-4}
Microwave	3×10^{8} to 3×10^{6}	3×10^{-2} to 3.0	3.98×10^{-4} to 3.98×10^{-2}
Far infrared	3×10^{6} to 3×10^{4}	3.0 to 3.0×10^{2}	3.98×10^{-2} to 3.98
Near infrared	3×10^{4} to 7.0×10^{2}	3.0×10^{2} to 1.5×10^{4}	3.98 to 1.71×10^{2}
Visible			
Red	7.0×10^{2}	1.43×10^{4}	1.71×10^{2}
Orange	6.2×10^{2}	1.61×10^{4}	1.93×10^{2}
Yellow	5.8×10^{2}	1.73×10^{4}	2.06×10^{2}
Green	5.3×10^{2}	1.89×10^{4}	2.26×10^{2}
Blue	4.7×10^{2}	2.13×10^{4}	2.54×10^{2}
Violet	4.2×10^{2}	2.38×10^{4}	2.85×10^{2}
Near ultraviolet	4.2×10^{2} to 3.0×10^{2}	2.4×10^{4} to 3.0×10^{4}	2.85×10^{2} to 3.99×10^{2}
Vacuum ultraviolet	3.0×10^{2} to 3	3.0×10^{4} to 3×10^{6}	3.99×10^{2} to 3.99×10^{4}
X-ray	3 to 3.0×10^{-3}	3.0×10^{6} to 3.0×10^{9}	3.99×10^{4} to 3.99×10^{7}

32 INFRARED ABSORPTION FREQUENCIES

A selection of the characteristic infrared absorption frequencies is tabulated for the types of organic compounds given in table 25, in the same order as in that table. Values are expressed as wavenumbers.

Type of compound	Vibration	Wavenumber range/cm^{-1}
Alkanes	C–H stretch C–H bend	2850–2960 1370–1480
Alkenes	C–H stretch C=C stretch C–H bend	3010–3090 1620–1680 900–995
Alkynes	C–H stretch C≡C stretch C–H bend	3300 2100–2260 610–680
Aromatics	C–H stretch C–C stretch C–H bend	3000–3100 1600 750
Halides	C–F stretch C–Cl stretch C–Br stretch C–I stretch	1000–1400 600–800 500–600 490–620
Nitro-	N=O stretch	1510–1565 1345–1385
Alcohols, phenols	O–H stretch O–H stretch (hydrogen bonded) O–H bend and C–O stretch (coupled)	3610–3640 3200–3600 1050–1410
Thiols	S–H stretch	2550–2600
Amines	N–H stretch N–H bend	3300–3500 1560–1650
Carboxylic acids	O–H stretch (hydrogen bonded) C=O stretch	2500–3000 1700–1725
Esters	C=O stretch	1735–1750
Aldehydes	C=O stretch	1720–1740
Ketones	C=O stretch	1705–1725
Ethers	C–O stretch	1070–1150

33 NMR CHEMICAL SHIFTS

A selection of the characteristic proton (^1H) and carbon (^{13}C) nuclear magnetic resonance chemical shifts is tabulated for the types of organic compounds given in table 25, in the same order as in that table. Values are expressed in parts per million downfield from the reference tetramethylsilane (TMS).

Type of compound	Proton[a]	δ/ppm	Carbon[a]	δ/ppm
Alkanes	RCH_3	0.9–1.0	RCH_3	5–30
	R_2CH_2	1.3–1.4	R_2CH_2	15–55
	R_3CH	1.5	R_3CH	25–60
Alkenes	C=C–H	4.5–6.0	C=C	80–150
	C=C–C–H	1.6–1.9		
Alkynes	C≡C–H	1.8–3.1	C≡C	70–80
Aromatics	Ar–H	6.0–9.0	ArH	110–140
	Ar–C–H	2.5–3.5		
Halides	X–C–H	3.5–4.4	C–X	0–95
Nitro-	$H–C–NO_2$	4.4–4.7	C–N	50–75
Alcohols	C–O–H	1–6	C–O	35–85
	O–C–H	3.3–4.5		
Phenols	Ar–O–H	4–12		
Thiols	R–S–H	1–2		
	Ar–S–H	3–4		
Amines	C–N–H	0.5–4.5	C–N	50–75
	N–C–H	2.3–2.8		
Carboxylic acids	RCOOH	9–13	RCOOH	165–185
Esters	H–C–O–COR	3.7–4.8	RCOOR	165–180
	H–C–COOR	2.0–2.5	O–C	40–75
Aldehydes	RCHO	9.4–10.0	RCHO	185–210
	ArCHO	9.7–10.5		
Ketones	H–C–CO–R	2.2–2.7	C=O	185–210
Ethers	H–C–O–R	3.3–3.7	C–O	40–85

[a] R = alkyl, Ar = aryl, X = F, Cl, Br or I.

34 CHEMICAL SHIFTS AND MULTIPLICITIES OF RESIDUAL PROTONS IN DEUTERATED NMR SOLVENTS[a]

Solvent[b]	δ_H (multiplicity)
acetic acid-d_4	11.53 (1), 2.03 (5)
acetone-d_6	2.04 (5)
acetonitrile-d_3	1.93 (5)
benzene-d_6	7.15 (broad)
chloroform-d_1	7.26 (1)
N,N-dimethylformamide-d_7	8.01 (broad), 2.91 (5), 2.74 (5),
dimethylsulfoxide-d_6	2.49 (5)
p-dioxane-d_8	3.53 (multiplet)
ethanol-d_6	5.19 (1), 3.55 (broad), 1.11 (multiplet)
methanol-d_4	4.78 (1), 3.30 (5)
dichloromethane-d_2	5.32 (3)
nitrobenzene-d_5	8.11 (broad), 7.67 (broad), 7.50 (broad)
nitromethane-d_3	4.33 (5)
pyridine-d_5	7.55 (broad), 7.19 (broad)
tetrahydrofuran-d_8	3.58 (broad), 1.73 (broad)
toluene-d_8	7.09 (multiplet), 7.00 (broad), 6.98 (multiplet), 2.09 (5)
trifluoroacetic acid-d_1	11.50 (1)

[a] The residual proton consists of one proton of each kind in an otherwise completely deuterated molecule. For example, deuterated acetic acid has two different kinds of residual protons: $CD_2H–COOD$ and $CD_3–COOH$. The CD_2H proton, coupled to two D nuclei is at δ 2.03 with a multiplicity of 5 ($2nI + 1$); the carboxylic acid proton is a singlet at δ 11.53.

[b] Purity (atom %D) up to 99.96% for several solvents

35 NMR CHEMICAL SHIFTS OF COMMON SOLVENTS IN A VARIETY OF DEUTERATED NMR SOLVENTS

^1H NMR DATA

	Proton	Multiplicity	$CDCl_3$	$(CD_3)_2CO$	$(CD_3)_2SO$	C_6D_6	CD_3CN	CD_3OD	D_2O
Solvent residual peak			7.26	2.05	2.50	7.16	1.94	3.31	4.79
H_2O		s	1.56	2.84	3.33	0.40	2.13	4.87	
acetic acid	CH_3	s	2.10	1.96	1.91	1.55	1.96	1.99	2.08
acetone	CH_3	s	2.17	2.09	2.09	1.55	2.08	2.15	2.22
acetonitrile	CH_3	s	2.10	2.05	2.07	1.55	1.96	2.03	2.06
benzene	CH	s	7.36	7.36	7.37	7.15	7.37	7.33	
tert-butanol	CH_3	s	1.28	1.18	1.11	1.05	1.16	1.40	1.24
	OH	s			4.19	1.55	2.18		
tert-butyl methyl ether	CCH_3	s	1.19	1.13	1.11	1.07	1.14	1.15	1.21
	OCH_3	s	3.22	3.13	3.08	3.04	3.13	3.20	3.22
chloroform	CH	s	7.26	8.02	8.32	6.15	7.58	7.90	
cyclohexane	CH_2	s	1.43	1.43	1.40	1.40	1.44	1.45	
1,2-dichloroethane	CH_2	s	3.73	3.87	3.90	2.90	3.81	3.78	
dichloromethane	CH_2	s	5.30	5.63	5.76	4.27	5.44	5.49	
diethyl ether	CH_3	t	1.21	1.11	1.09	1.11	1.12	1.18	1.17
	CH_2	q	3.48	3.41	3.38	3.26	3.42	3.49	3.56
diglyme	CH_2	m	3.65	3.56	3.51	3.46	3.53	3.61	3.67
	CH_2	m	3.57	3.47	3.38	3.34	3.45	3.58	3.61
	OCH_3	s	3.39	3.28	3.24	3.11	3.29	3.35	3.37
1,2-dimethoxyethane	CH_3	s	3.40	3.28	3.24	3.12	3.28	3.35	3.37
	CH_2	s	3.55	3.46	3.43	3.33	3.45	3.52	3.06
dimethylacetamide	CH_3CO	s	2.09	1.97	1.96	1.60	1.97	2.07	2.08
	NCH_3	s	3.02	3.00	2.94	2.57	2.96	3.31	3.06
	NCH_3	s	2.94	2.83	2.78	2.05	2.83	2.92	2.90

^{1}H NMR DATA *(continued)*

	Proton	Multiplicity	$CDCl_3$	$(CD_3)_2CO$	$(CD_3)_2SO$	C_6D_6	CD_3CN	CD_3OD	D_2O
dimethylformamide	CH	s	8.02	7.96	7.95	7.63	7.92	7.97	7.92
	CH_3	s	2.96	2.94	2.89	2.36	2.89	2.99	3.01
	CH_3	s	2.88	2.78	2.73	1.86	2.77	2.86	2.85
dimethyl sulfoxide	CH_3	s	2.62	2.52	2.54	1.68	2.50	2.65	2.71
dioxane	CH_2	s	3.71	3.59	3.57	3.35	3.60	3.66	3.75
ethanol	CH_3	t	1.25	1.12	1.06	0.96	1.12	1.19	1.17
	CH_2	q	3.72	3.57	3.44	3.34	3.54	3.60	3.65
	OH	s	1.32	3.39	4.63		2.47		
ethyl acetate	CH_3CO	s	2.05	1.97	1.99	1.65	1.97	2.01	2.07
	$\mathit{CH_2}CH_3$	q	4.12	4.05	4.03	3.89	4.06	4.09	4.14
	$CH_2\mathit{CH_3}$	t	1.26	1.20	1.17	0.92	1.20	1.24	1.24
ethyl methyl ketone	CH_3CO	s	2.14	2.07	2.07	1.58	2.06	2.12	2.19
	$\mathit{CH_2}CH_3$	q	2.46	2.45	2.43	1.81	2.43	2.50	3.18
	$CH_2\mathit{CH_3}$	t	1.06	0.96	0.91	0.85	0.96	1.01	1.26
hexane	CH_3	t	0.88	0.88	0.86	0.89	0.89	0.90	
	CH_2	m	1.26	1.28	1.25	1.24	1.28	1.29	
methanol	CH_3	s	3.49	3.31	3.16	3.07	3.28	3.34	3.34
	OH	s	1.09	3.12	4.01		2.16		
propan-2-ol	CH_3	d	1.22	1.10	1.04	0.95	1.09	1.50	1.17
	CH	sep	4.04	3.90	3.78	3.67	3.87	3.92	4.02
pyridine	CH(2)	m	8.62	8.58	8.58	8.53	8.57	8.53	8.52
	CH(3)	m	7.29	7.35	7.39	6.66	7.33	7.44	7.45
	CH(4)	m	7.68	7.76	7.79	6.98	7.73	7.85	7.87
tetrahydrofuran	CH_2	m	1.85	1.79	1.76	1.40	1.80	1.87	1.88
	CH_2O	m	3.76	3.63	3.60	3.57	3.64	3.71	3.74
toluene	CH_3	s	2.36	2.32	2.30	2.11	2.33	2.32	
	CH(*o*/*p*)	m	7.17	7.1–7.2	7.18	7.02	7.1–7.3	7.16	
	CH(*m*)	m	7.25	7.1–7.2	7.25	7.13	7.1–7.3	7.16	

	Carbon	$CDCl_3$	$(CD_3)_2CO$	$(CD_3)_2SO$	C_6D_6	CD_3CN	CD_3OD	D_2O
Solvent residual peak		77.16 ± 0.06	29.84 ± 0.01	39.52 ± 0.06	128.06 ± 0.02	1.32 ± 0.02	49.00 ± 0.01	
			206.26 ± .013			118.26 ± 0.02		
acetic acid	CO	175.99	172.31	171.93	175.82	173.21	175.11	177.21
	CH_3	20.81	20.51	20.95	20.37	20.73	20.56	21.03
acetone	CO	207.07	205.87	206.31	204.43	207.43	209.67	215.94
	CH_3	30.92	30.60	30.56	30.14	30.91	30.67	30.89
acetonitrile	CN	116.43	117.60	117.91	116.02	118.26	118.06	119.68
	CH_3	1.89	1.12	1.03	0.20	1.79	0.85	1.47
benzene	CH	128.37	129.15	128.30	128.62	129.32	129.34	
tert-butanol	C	69.15	68.13	66.88	68.19	68.74	69.40	70.36
	CH_3	31.25	30.72	30.38	30.47	30.68	30.91	30.29
tert-butyl methyl ether	OCH_3	49.45	49.35	48.70	49.19	49.52	49.66	49.37
	C	72.87	72.81	72.04	72.40	73.17	74.32	75.62
	CCH_3	26.99	27.24	26.79	27.09	27.28	27.22	26.60
chloroform	CH	77.36	79.19	79.16	77.79	79.17	79.44	
cyclohexane	CH_2	26.94	27.51	26.33	27.23	27.63	27.96	
1,2-dichloroethane	CH_2	43.50	45.25	45.02	43.59	45.54	45.11	
dichloromethane	CH_2	53.52	54.95	54.84	53.46	55.32	54.78	
diethyl ether	CH_3	15.20	15.78	15.12	15.46	15.63	15.46	14.77
	CH_2	65.91	66.12	62.05	65.94	66.32	66.88	66.42
diglyme	CH_3	59.01	58.77	57.98	58.66	58.90	59.06	58.67
	CH_2	70.51	71.03	69.54	70.87	70.99	71.33	70.05
	CH_2	71.90	72.63	71.25	72.35	72.63	72.92	71.63
1,2-dimethoxyethane	CH_3	59.08	58.45	58.01	58.68	58.89	59.06	58.67
	CH_2	71.84	72.47	17.07	72.21	72.47	72.72	71.49
dimethylacetamide	CH_3	21.53	21.51	21.29	21.16	21.76	21.32	21.09
	CO	171.07	170.61	169.54	169.95	171.31	173.32	174.57
	NCH_3	35.28	34.89	37.38	34.67	35.17	35.50	35.03
	NCH_3	38.13	37.92	34.42	37.03	38.26	38.43	38.76

^{13}C NMR DATA *(continued)*

	Carbon	$CDCl_3$	$(CD_3)_2CO$	$(CD_3)_2SO$	C_6D_6	CD_3CN	CD_3OD	D_2O
dimethylformamide	CH	162.62	162.79	162.29	162.13	163.31	164.73	165.53
	CH_3	36.50	36.15	35.73	35.25	36.57	36.89	37.54
	CH_3	31.45	31.03	30.73	30.72	31.32	31.61	32.03
dimethyl sulfoxide	CH_3	40.76	41.23	40.45	40.03	41.31	40.45	39.39
dioxane	CH_2	67.14	67.60	66.36	67.16	67.72	68.11	67.19
ethanol	CH_3	18.41	18.89	18.51	18.72	18.80	18.40	17.47
	CH_2	58.28	57.72	56.07	57.86	57.96	58.26	58.05
ethyl acetate	$\mathit{CH_3}CO$	21.04	20.83	20.68	20.56	21.16	20.88	21.15
	CO	171.36	170.96	170.31	170.44	171.68	172.89	175.26
	CH_2	60.49	60.56	59.74	60.21	60.98	61.50	62.32
	CH_3	14.19	14.50	14.40	14.19	14.54	14.49	13.92
ethyl methyl ketone	$\mathit{CH_3}CO$	29.49	29.30	29.26	28.56	29.60	29.39	29.49
	CO	209.56	208.30	208.72	206.55	209.88	212.16	218.43
	$\mathit{CH_2}CH_3$	36.89	36.75	35.83	36.36	37.09	37.34	37.27
	$CH_2\mathit{CH_3}$	7.86	8.03	7.61	7.91	8.14	8.09	7.87
hexane	CH_3	14.14	14.34	13.88	14.32	14.43	14.45	
	CH_2(2)	22.70	23.28	22.05	23.04	23.40	23.68	
	CH_2(3)	31.64	32.30	30.95	31.96	32.36	32.73	
methanol	CH_3	50.41	49.77	48.59	49.97	49.90	49.86	49.50
propan-2-ol	CH_3	25.14	25.67	25.43	25.18	25.55	25.27	24.38
	CH	64.50	63.85	64.92	64.23	64.30	64.71	64.88
pyridine	CH(2)	149.90	150.67	149.58	150.27	150.76	150.07	149.18
	CH(3)	123.75	124.57	123.84	123.58	127.76	125.53	125.12
	CH(4)	135.96	136.56	136.05	135.28	136.89	138.35	138.27
tetrahydrofuran	CH_2	25.62	26.15	25.14	25.72	26.27	26.48	25.67
	CH_2O	67.97	68.07	67.03	67.80	68.33	68.83	68.68
toluene	CH_3	21.46	21.46	20.99	21.10	21.50	21.50	
	C(*i*)	137.89	138.48	137.35	137.91	138.90	138.85	
	CH(*o*)	129.07	129.76	128.88	129.33	129.94	129.91	
	CH(*m*)	128.26	129.03	128.18	128.56	129.23	129.20	
	CH(*p*)	125.33	126.12	125.29	125.68	126.28	126.29	

36 IMPORTANT NMR-ACTIVE NUCLEI

In order for an atomic nucleus to be NMR-active, it must have a non-zero value of I, the nuclear spin. Listed below are some of the more commonly used NMR-active nuclei. Note that the list is not exhaustive.

Isotope	Abundance (%)	Relative receptivity compared to ^{1}H	Nuclear spin (I) (in multiples of $h/2\pi$)
^{1}H	99.9844	1.00	1/2
^{2}H (D)	0.0156	1.5×10^{-6}	1
^{6}Li	7.4	6.3×10^{-4}	1
^{7}Li	92.6	0.27	3/2
^{10}B	18.83	3.9×10^{-3}	3
^{11}B	81.17	0.13	3/2
^{13}C	1.108	1.8×10^{-4}	1/2
^{15}N	0.365	3.9×10^{-6}	1/2
^{19}F	100	0.83	1/2
^{23}Na	100	0.093	3/2
^{27}Al	100	0.21	5/2
^{29}Si	4.70	3.7×10^{-4}	1/2
^{31}P	100	0.066	1/2
^{51}V	100	0.38	7/2
^{59}Co	100	0.28	7/2
^{69}Ga	60.4	0.042	3/2
^{71}Ga	39.6	0.057	3/2
^{107}Ag	51.8	3.5×10^{-5}	1/2
^{109}Ag	48.2	4.9×10^{-5}	1/2
^{111}Cd	12.86	1.2×10^{-3}	1/2
^{113}Cd	12.34	1.3×10^{-3}	1/2
^{117}Sn	7.67	3.5×10^{-3}	1/2
^{119}Sn	8.68	4.5×10^{-3}	1/2
^{129}Xe	26.44	5.7×10^{-3}	1/2
^{195}Pt	33.7	3.4×10^{-3}	1/2

37 COMMON SINGLY CHARGED ($z = 1$) FRAGMENTS DETECTED BY MASS SPECTROMETRY

m/z	Small ion/neutral fragment
1	H
14	CH_2
15	CH_3
16	O, NH_2
17	OH
18	H_2O
19	F
20	HF
26	C_2H_2, CN
27	C_2H_3, HCN
28	C_2H_4, CO, H_2CN
29	C_2H_5, HCO
30	CH_2O, CH_2NH_2, NO
31	CH_2OH, CH_3NH_2
32	O_2, CH_4O, S
33	$CH_3 + H_2O$, HS
34	H_2S
35	Cl
36	HCl
37	H_2Cl
38	C_3H_2, C_2N, F_2
39	C_3H_3, HC_2N
41	C_3H_5, C_2H_3N
42	C_3H_6, NCO, C_2H_4N, $NCNH_2$

37 COMMON SINGLY CHARGED ($z = 1$) FRAGMENTS DETECTED BY MASS SPECTROMETRY *(continued)*

m/z	Small ion/neutral fragment
43	C_3H_7, CH_3CO, HCNO, $NHCH_2CH_2$
44	C_2H_4O, CO_2, N_2O, $CONH_2$
45	C_2H_5O, CO_2H, C_2H_7N
46	C_2H_6O, NO_2
47	CH_3S
48	CH_4S, SO, O_3
49	CH_2Cl
53	C_4H_5
54	C_3H_4N, C_4H_6
55	C_4H_7
56	C_4H_8
57	C_4H_9, C_3H_5O
59	C_3H_6OH
60	$C_2H_4O_2$
64	SO_2
65	C_5H_5
66	C_5H_6
67	C_5H_7
68	C_4H_6N
69	C_5H_9
70	C_5H_{10}
71	C_5H_{11}, C_4H_7O
76	C_6H_4
77	C_6H_5
78	C_6H_6
79	Br
80	HBr

38 DISSOCIATION CONSTANTS OF ACIDS AND HYDRATED METAL IONS

The dissociation constants are expressed as pK_a values ($pK_a = -\log_{10}K_a$) at 298 K, where all concentrations are expressed in mol L^{-1}.

For a monoprotic acid, K_a is the equilibrium constant for the equilibrium

$$HL(aq) + H_2O \rightleftharpoons H_3O^+(aq) + L^-(aq)$$

For a diprotic acid $H_2L(aq)$, $K_{a,1}$ is the equilibrium constant for the equilibrium

$$H_2L(aq) + H_2O \rightleftharpoons H_3O^+(aq) + HL^-(aq)$$

and $K_{a,2}$ is the equilibrium constant for the equilibrium

$$HL^-(aq) + H_2O \rightleftharpoons H_3O^+(aq) + L^{2-}(aq)$$

K_a values for triprotic and tetraprotic acids are defined in a similar manner.

For hydrated metal ions, $[M(H_2O)_n]^{m+}$, K_a is the equilibrium constant for the equilibrium

$$[M(H_2O)_n]^{m+} + H_2O \rightleftharpoons [M(H_2O)_{n-1}(OH)]^{(m-1)+} + H_3O^+(aq)$$

The dissociation constant of a base, B, is given in terms of the pK_a value of its conjugate acid $BH^+(aq)$

$$BH^+(aq) + H_2O \rightleftharpoons B(aq) + H_3O^+(aq)$$

The pK_b for a base may be calculated from the pK_a value of its conjugate acid from the expression

$$pK_w = pK_a + pK_b$$

where $K_w = [H_3O^+][OH^-] = 1.0 \times 10^{-14}$ at 298 K and hence $pK_w = 14.00$ at 298 K.

Note that K_a and pK_a may be interconverted as follows. Consider the equilibrium

$$CH_3COOH + H_2O \rightleftharpoons CH_3COO^- + H_3O^+$$

K_a for acetic acid (CH_3COOH) = 1.74×10^{-5}, so

$$K_a = \frac{[CH_3COO^-][H_3O^+]}{[CH_3COOH]} = 1.74 \times 10^{-5}$$

$pK_a = -\log_{10}(1.74 \times 10^{-5}) = 4.759$ and $K_a = 10^{(-4.759)} = 1.74 \times 10^{-5}$.

Monoprotic acids[a]

Name	Formula	pK_a
Acetic acid	CH_3COOH	4.76
Ammonium ion	NH_4^+	9.24
Anilinium ion	$C_6H_5NH_3^+$	4.60
Arsenious acid	H_3AsO_3	9.29
Benzoic acid	C_6H_5COOH	4.20
Boric acid	H_3BO_3	9.24
Butanoic acid	$CH_3(CH_2)_2COOH$	4.82
Chlorous acid	$HClO_2$	1.95
Ethylammonium ion	$C_2H_5NH_3^+$	10.64
Formic acid	$HCOOH$	3.74
Hydrocyanic acid	HCN	9.21
Hydrofluoric acid	HF	3.17
Hydrazoic acid	HN_3	4.65
Hydrogen peroxide	H_2O_2	11.65
Hydroxylammonium ion	NH_3OH^+	5.96
Hypobromous acid	$HOBr$	8.63
Hypochlorous acid	$HOCl$	7.53
Hypoiodous acid	HOI	10.64
Hypophosphorous acid	H_3PO_2	1.3
Methylammonium ion	$CH_3NH_3^+$	10.64
Nitrous acid	HNO_2	3.15
Nitric acid	HNO_3	-1.30
Phenol	C_6H_5OH	9.98
Propanoic acid	CH_3CH_2COOH	4.87
Pyridinium ion	$C_5H_5NH^+$	5.23

[a] For a more complete set of pK_a values of organic acids, see table 25: 'Properties of organic compounds'.

Diprotic acids[a]

Name	Formula	$pK_{a,1}$	$pK_{a,2}$
Carbonic acid	$CO_2(aq)$	6.35	10.33
Chromic acid	H_2CrO_4	–0.2	6.50
Hydrazinium (2+) ion	$N_2H_6^{2+}$	–0.9	7.98
Hydrogen sulfide	H_2S	7.02	13.9
Malonic acid	$COOHCH_2COOH$	2.85	5.70
Oxalic acid	$COOHCOOH$	1.25	4.27

Diprotic acids[(a)] *(continued)*

Name	Formula	p$K_{a,1}$	p$K_{a,2}$
Phosphorous acid	H_3PO_3	1.5	6.78
Salicylic acid	OHC_6H_4COOH	2.97 (RCOOH)	13.70 (ROH)
Selenous acid	H_2SeO_3	2.65	7.9
Selenic acid	H_2SeO_4	<0	1.70
Silicic acid	H_4SiO_4	9.84	13.2
Succinic acid	$COOH(CH_2)_2COOH$	4.21	5.64
Sulfurous acid	SO_2(aq)	1.86	7.19
Sulfuric acid	H_2SO_4	<0	1.99
Thiosulfuric acid	$H_2S_2O_3$	0.6	1.6

[(a)] For a more complete set of pK_a values of organic acids, see table 25: 'Properties of organic compounds'.

Triprotic acids

Name	Formula	p$K_{a,1}$	p$K_{a,2}$	p$K_{a,3}$
Arsenic acid	H_3AsO_4	2.24	6.96	11.50
Phosphoric acid	H_3PO_4	2.15	7.20	12.38

Tetraprotic acids

Name	Formula	p$K_{a,1}$	p$K_{a,2}$	p$K_{a,3}$	p$K_{a,4}$
Pyrophosphoric acid	$H_4P_2O_7$	0.9	2.31	6.69	9.42

Hydrated metal ions

Name	pK_a
Aluminium(III) ion	4.96
Beryllium(II) ion	6.5
Chromium(III) ion	3.95
Copper(II) ion	7.34
Iron(III) ion	2.17
Lead(II) ion	7.9
Manganese(II) ion	10.59
Nickel(II) ion	9.86
Scandium(III) ion	4.22
Uranium(IV) ion	0.68
Vanadium(II) ion	2.9
Zinc(II) ion	8.96

39 AQUEOUS CONCENTRATIONS OF COMMON ACIDS

Reagent	Approximate				Volume required to make 1 L of 1 M solution
	Wt%	Molar mass	Density/ g mL^{-1}	Molarity/M	
Ethanoic acid (acetic acid) (CH_3COOH)	99.5	60.05	1.05	17.4	58
Formic acid (HCOOH)	90	46.02	1.20	23.4	43
Hydrobromic acid (HBr)	48	80.92	1.50	8.89	112
Hydrochloric acid (HCl)	37	36.5	1.18	11.6	89
Hydrofluoric acid (HF)	46	20.01	1.15	32.1	38
Hydroiodic acid (HI)	57	127.9	1.70	7.57	132
Nitric acid (HNO_3)	70	63.02	1.42	16.0	63
Perchloric acid ($HClO_4$)	70	100.5	1.66	11.6	86
Phosphoric acid (H_3PO_4)	85	98	1.69	14.7	69
Sulfuric acid (H_2SO_4)	96	98.1	1.84	18.0	56
Trifluoroacetic acid (CF_3COOH)	99	114.02	1.489	12.9	77
Trifluoromethanesulfonic acid (triflic acid) (CF_3SO_3H)	99	150.08	1.70	11.3	89

40 COMMON ACID–BASE INDICATORS

pH range and colours			Indicator	Composition per 10 mL	pK_a
Red	0.2 – 1.8	Yellow	Cresol red	0.1% in aqueous solution	–
Red	1.2 – 2.8	Yellow	Thymol Blue	1–2 drops 0.1% solution in aqueous solution	1.7
Colourless	2.4 – 4.0	Yellow	2,4-Dinitrophenol	1–2 drops 0.1% solution in 50% ethanol	4.1
Red	2.9 – 4.0	Yellow	Methyl yellow	1 drop 0.1% solution in 90% ethanol	3.1
Red	3.1 – 4.4	Orange	Methyl orange	1 drop 0.1% aqueous solution	3.7
Yellow	3.0 – 4.6	Blue-violet	Bromophenol blue	1 drop 0.1% aqueous solution	4.2
Yellow	3.7 – 5.2	Violet	Alizarin sodium sulfonate	1 drop 0.1% aqueous solution	4.5
Yellow	4.0 – 5.6	Blue	Bromocresol green	1 drop 0.1% aqueous solution	4.7
Red	4.4 – 6.2	Yellow	Methyl red	1 drop 0.1% aqueous solution	5.1
Yellow	5.2 – 6.8	Purple	Bromocresol purple	1 drop 0.1% aqueous solution	6.3
Colourless	5.0 – 7.0	Yellow	*p*-Nitrophenol	1–5 drops 0.1% aqueous solution	7.2
Red	5.0 – 8.0	Blue	Azolitmin (litmus)	5 drops 0.5% aqueous solution	–
Yellow	6.0 – 7.6	Blue	Bromothymol blue	0.1% in aqueous solution	7.0
Yellow	6.4 – 8.0	Red	Phenol red	1 drop 0.1% aqueous solution	7.9
Yellow	7.2 – 8.8	Red	Cresol red	1 drop 0.1% aqueous solution	8.3
Yellow	8.0 – 9.6	Blue	Thymol blue	1–5 drops 0.1% aqueous solution	8.9
Colourless	8.0 – 10.0	Red	Phenolphthalein	1–5 drops 0.1% solution in 70% ethanol	9.6
Colourless	9.4 – 10.6	Blue	Thymolphthalein	1 drop 0.1% solution in 90% ethanol	9.2
Blue	10.1 – 11.1	Red	Nile blue	1 drop 0.1% aqueous solution	9.7

41 COMMON BUFFERS

Buffering system	Useful buffering pH range at 25 °C
Glycine/hydrochloric acid (NH_2CH_2COOH/HCl)	2.2–3.6
Potassium hydrogen phthalate/hydrochloric acid ($KC_8H_5O_4/HCl$)	2.2–4.0
Citric acid/sodium citrate ($C_6H_8O_7/Na_3C_6H_5O_7$)	3.0–6.2
Sodium acetate/acetic acid ($NaCH_3COO/CH_3COOH$)	3.7–5.6
Potassium hydrogen phthalate/sodium hydroxide ($KC_8H_5O_4/NaOH$)	4.1–5.9
Potassium dihydrogen orthophosphate/sodium hydroxide ($KH_2PO_4/NaOH$)	5.8–8.0
Sodium tetraborate/hydrochloric acid ($Na_2B_4O_7/HCl$)	8.1–9.2
Glycine/sodium hydroxide ($NH_2CH_2COOH/NaOH$)	8.6–10.6
Sodium carbonate/sodium hydrogen carbonate ($Na_2CO_3/NaHCO_3$)	9.2–10.8
Sodium hydrogen carbonate/sodium hydroxide ($NaHCO_3/NaOH$)	9.60–11.0

42 BUFFERING RANGES OF SOME COMMON BIOLOGICAL BUFFERS (0.1 M)

Buffer	Name and structure	Useful pH range	pK_a (at 298 K)
MES	2-(*N*-morpholino)ethanesulfonic acid	5.5–6.7	6.10
BIS-TRIS	bis(2-hydroxyethyl)amino-tris(hydroxymethyl)methane	5.8–7.2	6.50
ADA	*N*-(2-acetamido)iminodiacetic acid	6.0–7.2	6.59
ACES	*N*-(2-acetamido)-2-aminoethanesulfonic acid	6.1–7.5	6.78
PIPES	piperazine-*N*,*N*'-bis(2-ethanesulfonic acid)	6.1–7.5	6.76
MOPSO	3-(*N*-morpholinyl)-2-hydroxypropanesulfonic acid sodium salt	6.2–7.6	6.90
BIS-TRIS PROPANE	1,3-bis(tris(hydroxymethyl)methylamino)propane	6.3–9.5	6.8, 9.0
BES	*N*,*N*-bis(2-hydroxyethyl)-2-aminoethanesulfonic acid	6.4–7.8	7.09
MOPS	3-(*N*-morpholino)propanesulfonic acid	6.5–7.9	7.20
TES	2-[[1,3-dihydroxy-2-(hydroxymethyl)propan-2-yl]amino]ethanesulfonic acid	6.8–8.2	7.40

Buffer	Name and structure	Useful pH range	pK_a (at 298 K)
HEPES	4-(2-hydroxyethyl)-1-piperazineethanesulfonic acid 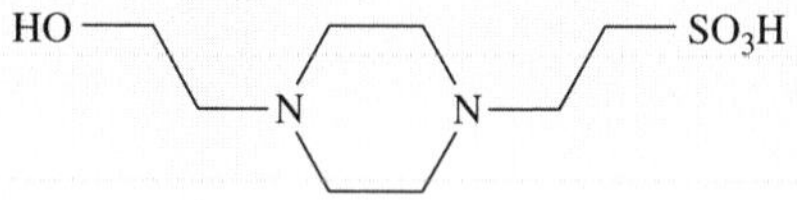	6.8–8.2	7.48
DIPSO	*N,N*-bis(2-hydroxyethyl)-3-amino-2-hydroxypropanesulfonic acid 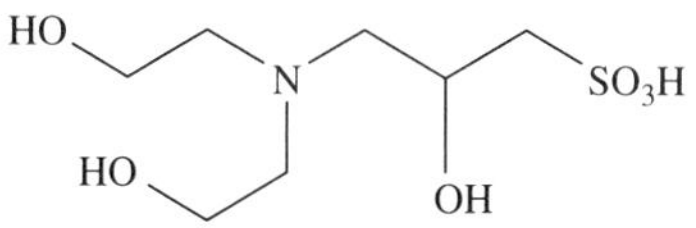	7.0–8.2	7.60
MOBS	4-(*N*-morpholino)butanesulfonic acid 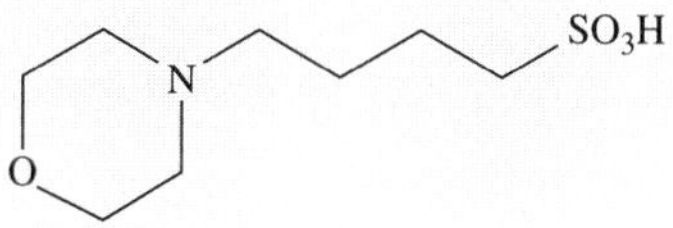	6.9–8.3	7.60
TAPSO	3-[[1,3-dihydroxy-2-(hydroxymethyl)propan-2-yl] amino]-2-hydroxypropane-1-sulfonic acid 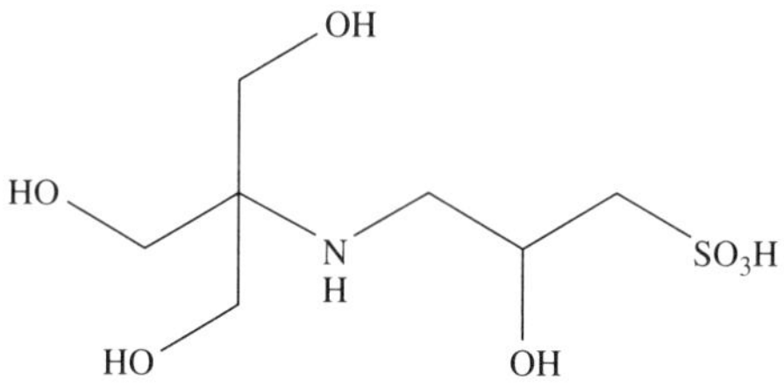	7.0–8.2	7.60
TRIS	Tris(hydroxymethyl)aminomethane 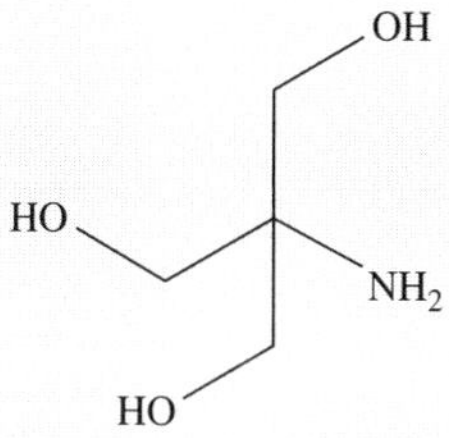	7.0–9.0	8.06

Buffer	Name and structure	Useful pH range	pK_a (at 298 K)
HEPPSO	4-(2-hydroxyethyl)-piperazine-1-(2-hydroxy)-propanesulfonic acid 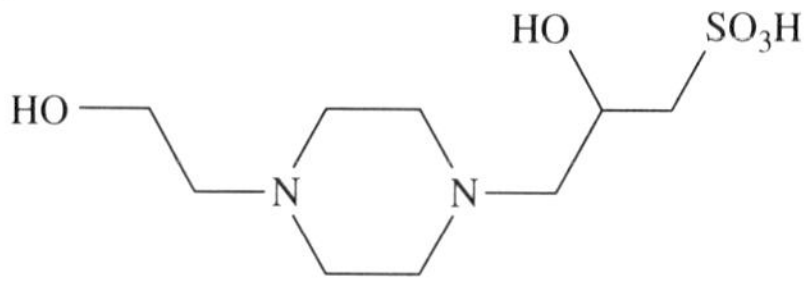	7.1–8.5	7.80
POPSO	Piperazine-1,4-bis(2-hydroxy-3-propanesulfonic acid) dihydrate 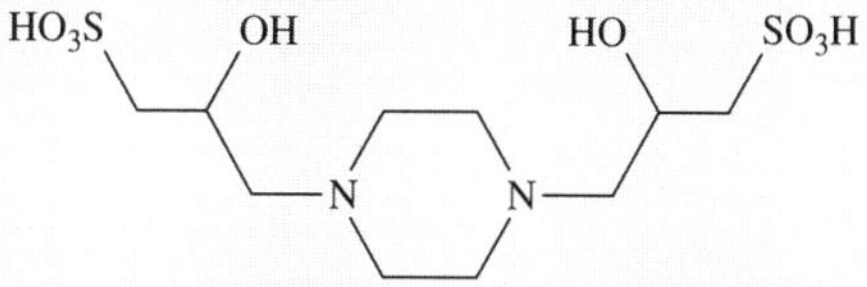	7.2–8.5	7.80
EPPS	4-(2-hydroxyethyl)piperazine-1-propanesulfonic acid 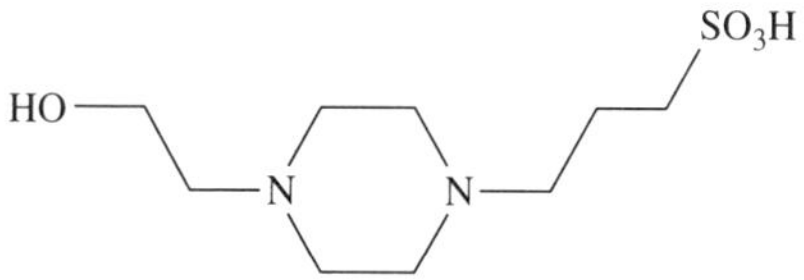	7.3–8.7	8.00
TRICINE	N-[tris(hydroxymethyl)methyl]-glycine 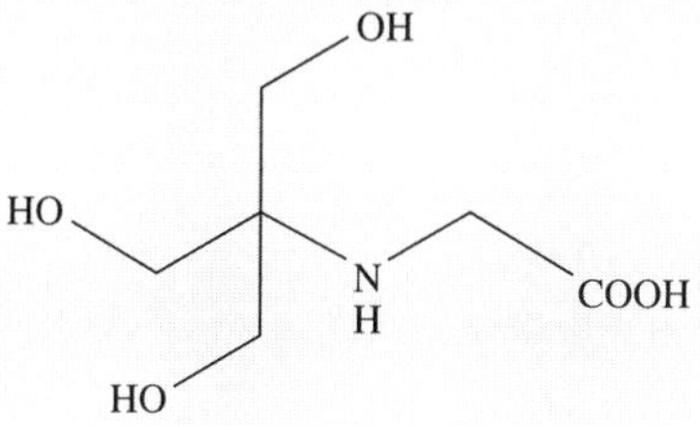	7.4–8.8	8.05
BICINE	*N,N*-bis(2-hydroxyethyl)glycine 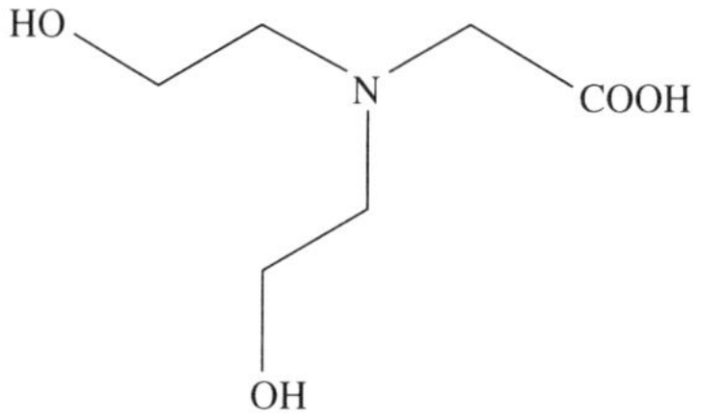	7.6–9.0	8.26

42 BUFFERING RANGES OF SOME COMMON BIOLOGICAL BUFFERS (0.1 M) *(continued)*

Buffer	Name and structure	Useful pH range	pK_a (at 298 K)
HEPBS	*N*-(2-hydroxyethyl)piperazine-*N*′-(4-butanesulfonic acid) 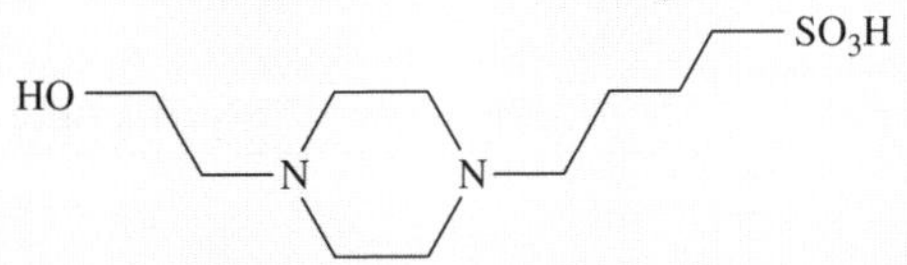	7.6–9.0	8.30
TAPS	3-[[1,3-dihydroxy-2-(hydroxymethyl)propan-2-yl] amino]propane-1-sulfonic acid 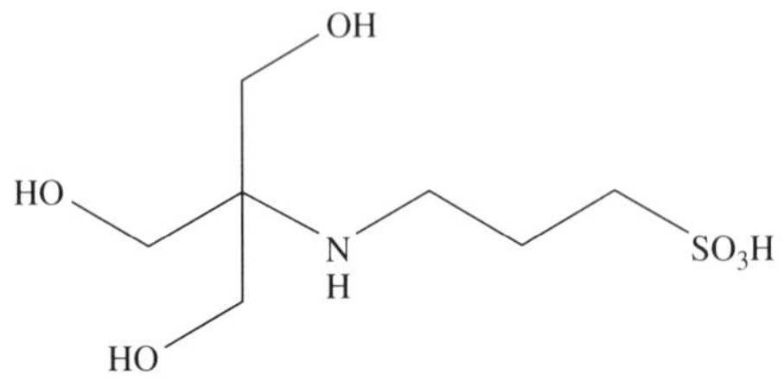	7.7-9.1	8.40
AMPD	2-amino-2-methyl-1,3-propanediol 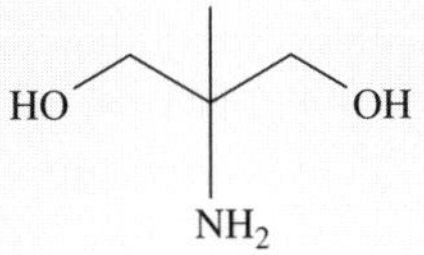	7.8–9.7	8.80
CHES	*N*-cyclohexyl-2-aminoethanesulfonic acid 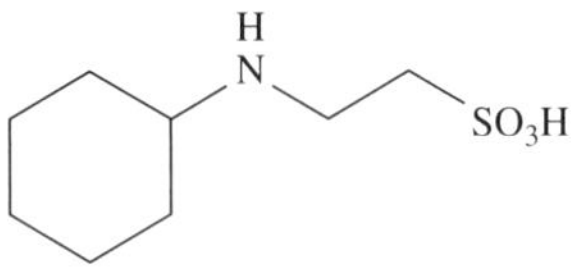	8.6–10.0	9.49
CAPSO	3-(cyclohexylamino)-2-hydroxypropane-1-sulfonic acid 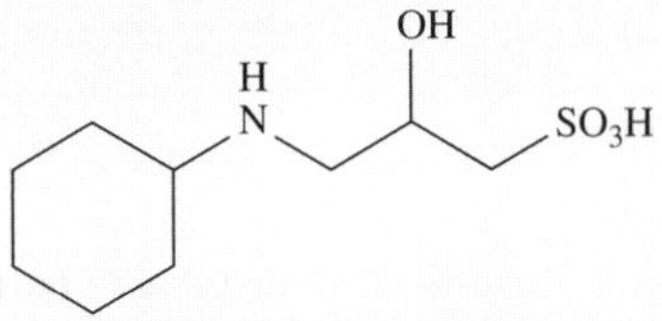	8.9–10.3	9.60
AMP	2-amino-2-methylpropan-1-ol 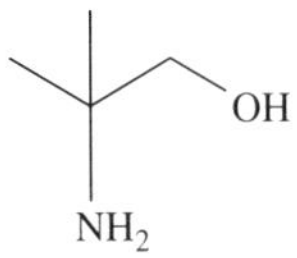	9.0–10.5	9.70
CAPS	*N*-cyclohexyl-3-aminopropanesulfonic acid 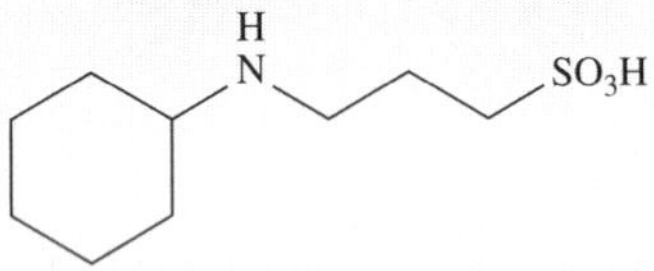	9.7–11.1	10.40

43 IONIC PROPERTIES OF WATER

The electrical conductivity (κ) is defined for a unit cube of water when there is unit potential difference between opposite sides.
The autoprotolysis constant (ionic product) of water, $K_w = [H_3O^+][OH^-]$, is given with concentrations expressed in mol L^{-1}.

Property	Temperature/°C							
	0	10	20	25	30	40	50	100
Electrical conductivity κ/μS m^{-1}	1.2	2.3	4.2	5.5	7.1	11.3	17.1	
Ionic product $10^{14}K_w$	0.114	0.292	0.681	1.008	1.47	2.92	5.5	55

44 MOLAR CONDUCTIVITIES OF AQUEOUS SOLUTIONS AT 298 K

Molar conductivity (Λ) is the electrical conductivity of a solution containing 1 mole placed between electrodes 1 metre apart. All data refer to 298 K.

Solute	Λ/mS m^2 mol^{-1} Concentration/mol L^{-1}			
	1.0	0.1	0.01	0.001
$AgNO_3$	7.78	10.91	12.48	13.05
$BaCl_2$	13.80	19.74	23.85	26.45
CH_3COOH		0.52	1.6	4.8
HBr	33.45	39.19	41.37	42.29
HCl	33.22	39.11	41.19	42.12
HF	2.43	3.91	9.61	
HI	34.39	39.40	41.28	42.17
HNO_3		38.50	40.60	
H_2SO_4		46.86	61.60	78.16
KBr	11.72	13.12	14.32	14.88
KCl	11.19	12.90	14.13	14.71
KI	11.8	13.06	14.22	14.73
KNO_3	9.25	12.04	13.58	14.18
NH_4Cl	10.7	12.88	14.13	14.76
NaCl	8.58	10.67	11.85	12.37
NaOH		21.82	23.80	24.47
$NaOOCCH_3$	4.92	7.28	8.38	8.85
Na_2SO_4		16.3	21.36	24.46

45 IONIC MOLAR CONDUCTIVITIES AT INFINITE DILUTION AT 298 K

Ion	Λ_+/mS m^2 mol^{-1}	Ion	Λ_-/mS m^2 mol^{-1}
Ag^+	6.19	Br^-	7.81
Al^{3+}	18.3	$CH_3CO_2^-$	4.09
Ba^{2+}	12.73	Cl^-	7.64
Ca^{2+}	11.90	ClO_3^-	6.46
Cd^{2+}	10.8	ClO_4^-	6.74
Co^{2+}	11	CN^-	8.2
Cr^{3+}	20.1	CO_3^{2-}	11.86
Cs^+	7.73	F^-	5.54
Cu^{2+}	10.7	$Fe(CN)_6^{4-}$	44.2
Fe^{2+}	10.8	HCO_3^-	4.45
Fe^{3+}	20.5	I^-	7.68
H^+	35.01	NO_3^-	7.15
Hg^{2+}	12.72	OH^-	19.92
K^+	7.35	SO_4^{2-}	16.00
Li^+	3.87		
Mg^{2+}	10.61		
Mn^{2+}	10.7		
Na^+	5.01		
NH_4^+	7.35		
Ni^{2+}	10		
Pb^{2+}	14.0		
Rb^+	7.78		
Sr^{2+}	11.89		
Zn^{2+}	10.56		

46 SOLUBILITY OF GASES IN WATER

Solubility is expressed as g gas per kg water at a total pressure of 101.325 kPa (that is, the sum of the partial pressure of the gas plus the vapour pressure of water at the given temperature).

	Solubility/g kg^{-1}							
	Temperature/°C							
Gas	0	20	25	30	35	40	60	80
Ar	0.099	0.059	0.053	0.049	0.046	0.042	0.030	
CH_4	0.0396	0.0232	0.0209	0.0191	0.0173	0.0159	0.0114	0.0070
C_2H_4	0.281	0.149	0.131	0.118				
C_2H_6	0.132	0.0620	0.0535	0.049	0.041	0.0366	0.0239	0.0134
CO	0.0440	0.0284	0.0260	0.024	0.022	0.0208	0.0152	0.0098
CO_2	3.35	1.69	1.45	1.26	1.10	0.973	0.576	
Cl_2	5.0	7.29	6.41	5.72	5.10	4.59	3.30	2.23
H_2	0.00192	0.00160	0.00154	0.00147	0.00142	0.00138	0.00118	0.00079
H_2S	7.07	3.85	3.38	2.98	2.65	2.36	1.48	0.77
He	0.0017	0.0015	0.0015	0.0014	0.0014	0.0014	0.0013	
N_2	0.0294	0.0190	0.0175	0.0162	0.0150	0.0139	0.0105	0.0066
NH_3	897	529	480	410		316	168	65
O_2	0.0694	0.0434	0.0393	0.0359	0.0332	0.0308	0.0227	0.0138
SO_2	228	113	94.1	78.0	64.7	54.1		

47 VAPOUR PRESSURE AND DENSITY OF WATER AND OF MERCURY AT DIFFERENT TEMPERATURES

	Water			Mercury		
t/°C	P/Pa	P/mmHg	ρ/g cm^{-3}	P/Pa	P/mmHg	ρ/g cm^{-3}
−10	286(l)	2.15(l)	0.9981(l)	0.008	0.0001	13.62
	260(s)	1.95(s)	0.9182(s)			
0	610(l)	4.58(l)	0.9998(l)	0.025	0.0002	13.60
	610(s)	4.58(s)	0.9168(s)			
5	872	6.54	1.0000	0.04	0.0003	13.58
10	1228	9.21	0.9997	0.07	0.0005	13.57
15	1705	12.79	0.9991	0.10	0.0008	13.56
20	2338	17.53	0.9982	0.16	0.0012	13.54
22	2643	19.83	0.9978	0.19	0.0014	13.54
24	2983	22.38	0.9973	0.22	0.0017	13.54
25	3167	23.76	0.9970	0.25	0.0018	13.53
26	3361	25.21	0.9968	0.27	0.0020	13.53
28	3780	28.35	0.9962	0.31	0.0024	13.53
30	4243	31.82	0.9956	0.37	0.0028	13.52
35	5623	42.18	0.9940	0.55	0.0041	13.51
50	1.233×10^4	92.51	0.9880	1.7	0.013	13.47
75	3.854×10^4	289.10	0.9749	8.8	0.066	13.41
100	1.013×10^5[(a)]	760.00[(a)]	0.9583	36.4	0.27	13.35
200	1.555×10^6	1.17×10^4	0.865(l)[(b)]	2.305×10^3	17.3	13.11
356.7	1.80×10^7	1.35×10^5	0.57(l)[(b)]	1.013×10^5	760	12.74
374.2	2.21×10^7	1.66×10^5	0.325(l)[(b)]	1.376×10^5	1032	12.70(l)[(b)]
			0.325(g)[(b)]			

[(a)] Near 100 °C the boiling point of water changes by about 0.28 °C/kPa (0.37 °C/10 mmHg) change in atmospheric pressure.

[(b)] Pressures greater than 101.325 kPa.

48 DENSITIES OF AQUEOUS SOLUTIONS AT 298 K

The density is expressed as g cm^{-3} at 298 K (unless otherwise indicated by a superscript). On this scale the density of water is 0.9970 g cm^{-3}.

Solute	Density/g cm^{-3}									
	Percentage by weight									
	10	20	30	40	50	60	70	80	90	100
$AgNO_3$	1.0866	1.1918	1.3185	1.469	1.659	1.912	2.228			
CH_3OH	0.9833	0.9723	0.9585	0.9466	0.9301	0.9099	0.8898	0.8643	0.8331	0.7865
C_2H_5OH	0.9804	0.9664	0.9507	0.9315	0.9098	0.8870	0.8634	0.8391	0.8136	0.7851
CH_3COOH	1.0107	1.0235	1.0350	1.0450	1.0534	1.0597	1.0637	1.0647	1.0605	1.0440
$C_{12}H_{22}O_{11}$ (sucrose)	1.0368	1.0794	1.1252	1.1744	1.2273	1.2890	1.3445			
HCl	1.0457	1.0957	1.1465							
HNO_3	1.0523	1.1123	1.1763	1.2417	1.3043	1.3600	1.4061	1.4439	1.4741	1.5040
H_2SO_4	1.0640	1.1365	1.2150	1.2991	1.3911	1.4940	1.6059	1.7221	1.8091	1.8255
KCl	1.0617	1.1307								
NH_3	0.956	0.918	0.889		0.832[a]	0.796[a]	0.755[a]	0.711[a]	0.665[a]	0.618[a]
NaCl	1.0688	1.1453								
NaOH	$1.1089^{293\ K}$	$1.2191^{293\ K}$	$1.3279^{293\ K}$	$1.4300^{293\ K}$	$1.5253^{293\ K}$					

[a] Determined in sealed tubes at 288 K.

49 STANDARD ELECTRODE POTENTIALS AND REDOX EQUILIBRIA IN AQUEOUS SOLUTION

The standard electrode potential, $E^{\ominus}$, of the half cell system

$$M^{z+} + ze^- \rightleftarrows M(s)$$

is the potential of the right-hand electrode minus that of the left-hand electrode in the cell

$$Pt, H_2|H^+ \parallel M^{z+}|M$$

when the activities of all ions in the cell are unity, when gases are at the standard state pressure of 10^5Pa and a temperature of 298 K and solids are in their most stable states under these standard state conditions.

The overall cell reaction may be represented as

$$M^{z+} + \frac{z}{2}H_{2(g)} \rightleftarrows M_{(s)} + zH^+$$

If the electrode potential is positive, the direction of spontaneous change is from left to right. If the electrode potential is negative, the direction of spontaneous change is from right to left.

When two half-cell equations are combined to give a third, the value of z, the number of electrons transferred in each half-cell equation, must be taken into account. Therefore, if equations 1 and 2 add to give equation 3, then

$$z_3E_3 = z_1E_1 + z_2E_2$$

For example:

$Au^{3+} + 2e^- \longrightarrow Au^+$ $\quad E_1 = 1.41\,V$

$Au^+ + e^- \longrightarrow Au(s)$ $\quad E_2 = 1.69\,V$

$Au^{3+} + 3e^- \longrightarrow Au(s)$ $\quad E_3 = \frac{1}{2}((2 \times 1.41) + 1.69)\,V = 1.50\,V$

Element	Reaction	$E^{\ominus}$/V
Aluminium	$Al^{3+} + 3e^- \rightleftarrows Al(s)$	−1.68
Antimony	$Sb_2O_5(s) + 4H^+ + 4e^- \rightleftarrows Sb_2O_3(s) + 2H_2O$	0.55
	$Sb_2O_3(s) + 6H^+ + 6e^- \rightleftarrows 2Sb(s) + 3H_2O$	0.15
	$SbO^+ + 2H^+ + 3e^- \rightleftarrows Sb(s) + H_2O$	0.20
	$Sb(s) + 3H^+ + 3e^- \rightleftarrows SbH_3(g)$	−0.51

Element	Reaction	$E^{\ominus}$/V
Arsenic	$H_3AsO_4 + 2H^+ + 2e^- \rightleftarrows H_3AsO_3 + H_2O$	0.57
	$H_3AsO_3 + 3H^+ + 3e^- \rightleftarrows As(s) + 3H_2O$	0.25
	$As(s) + 3H^+ + 3e^- \rightleftarrows AsH_3(g)$	−0.24
Barium	$Ba^{2+} + 2e^- \rightleftarrows Ba(s)$	−2.91
Beryllium	$Be^{2+} + 2e^- \rightleftarrows Be(s)$	−1.97
Bismuth	$BiO^+ + 2H^+ + 3e^- \rightleftarrows Bi(s) + H_2O$	0.32
	$BiOCl(s) + 2H^+ + 3e^- \rightleftarrows Bi(s) + H_2O + Cl^-$	0.16
	$Bi(OH)_6^- + 6H^+ + 2e^- \rightleftarrows Bi^{3+} + 6H_2O$	2.0
Boron	$H_3BO_3 + 3H^+ + 3e^- \rightleftarrows B(s) + 3H_2O$	−0.89
Bromine	$2BrO_3^- + 12H^+ + 10e^- \rightleftarrows Br_2(l) + 6H_2O$	1.51
	$2HOBr + 2H^+ + 2e^- \rightleftarrows Br_2(l) + 2H_2O$	1.60
	$Br_2(l) + 2e^- \rightleftarrows 2Br^-$	1.08
	$Br_2(aq) + 2e^- \rightleftarrows 2Br^-$	1.10
Cadmium	$Cd^{2+} + 2e^- \rightleftarrows Cd(s)$	−0.40
Caesium	$Cs^+ + e^- \rightleftarrows Cs(s)$	−3.02
Calcium	$Ca^{2+} + 2e^- \rightleftarrows Ca(s)$	−2.87
Carbon	$CO_2(g) + 2H^+ + 2e^- \rightleftarrows CO(g) + H_2O$	−0.10
	$CO_2(g) + 2H^+ + 2e^- \rightleftarrows HCOOH$	−0.11
	$2CO_2(g) + 2H^+ + 2e^- \rightleftarrows H_2C_2O_4$	−0.43
Cerium	$Ce^{4+} + e^- \rightleftarrows Ce^{3+}$	1.72
	$Ce^{3+} + 3e^- \rightleftarrows Ce(s)$	−2.34
Chlorine	$ClO_4^- + 2H^+ + 2e^- \rightleftarrows ClO_3^- + H_2O$	1.23
	$ClO_3^- + 2H^+ + e^- \rightleftarrows ClO_2(g) + H_2O$	1.13
	$ClO_3^- + 3H^+ + 2e^- \rightleftarrows HClO_2 + H_2O$	1.16
	$2ClO_3^- + 12H^+ + 10e^- \rightleftarrows Cl_2(g) + 6H_2O$	1.46
	$2HOCl + 2H^+ + 2e^- \rightleftarrows Cl_2(g) + 2H_2O$	1.63
	$Cl_2(g) + 2e^- \rightleftarrows 2Cl^-$	1.36
	$Cl_2(aq) + 2e^- \rightleftarrows 2Cl^-$	1.40

49 STANDARD ELECTRODE POTENTIALS *(continued)*

Element	Reaction	$E^{\ominus}$/V
Chromium	$Cr_2O_7^{2-} + 14H^+ + 6e^- \rightleftarrows 2Cr^{3+} + 7H_2O$	1.36
	$HCrO_4^- + 7H^+ + 3e^- \rightleftarrows Cr^{3+} + 4H_2O$	1.37
	$Cr^{3+} + e^- \rightleftarrows Cr^{2+}$	−0.42
	$Cr^{3+} + 3e^- \rightleftarrows Cr(s)$	−0.74
	$Cr^{2+} + 2e^- \rightleftarrows Cr(s)$	−0.90
Cobalt	$Co^{3+} + e^- \rightleftarrows Co^{2+}$	1.92
	$Co(NH_3)_6^{3+} + e^- \rightleftarrows Co(NH_3)_6^{2+}$	0.1
	$Co^{2+} + 2e^- \rightleftarrows Co(s)$	−0.28
	$Co(NH_3)_6^{2+} + 2e^- \rightleftarrows Co(s) + 6NH_3$	−0.41
	$Co(en)_3^{3+} + e^- \rightleftarrows Co(en)_3^{2+}$ [a]	−0.25
Copper	$Cu^{2+} + e^- \rightleftarrows Cu^+$	0.16
	$Cu(NH_3)_4^{2+} + e^- \rightleftarrows Cu(NH_3)_2^+ + 2NH_3$	0.09
	$Cu^{2+} + Cl^- + e^- \rightleftarrows CuCl(s)$	0.56
	$Cu^{2+} + Br^- + e^- \rightleftarrows CuBr(s)$	0.65
	$Cu^{2+} + I^- + e^- \rightleftarrows CuI(s)$	0.87
	$Cu^{2+} + 2e^- \rightleftarrows Cu(s)$	0.34
	$Cu(NH_3)_4^{2+} + 2e^- \rightleftarrows Cu(s) + 4NH_3$	−0.01
	$Cu^+ + e^- \rightleftarrows Cu(s)$	0.52
	$Cu(NH_3)_2^+ + e^- \rightleftarrows Cu(s) + 2NH_3$	−0.11
	$CuCl(s) + e^- \rightleftarrows Cu(s) + Cl^-$	0.12
Fluorine	$F_2(g) + 2e^- \rightleftarrows 2F^-$	2.89

[a] en = ethylenediamine.

Element	Reaction	$E^{\ominus}$/V
Gallium	$Ga^{3+} + 2e^- \rightleftarrows Ga^+$	−0.7
	$Ga^{3+} + 3e^- \rightleftarrows Ga(s)$	−0.55
Germanium	$Ge^{2+} + 2e^- \rightleftarrows Ge(s)$	0.24
Gold	$Au^{3+} + 2e^- \rightleftarrows Au^+$	1.41
	$Au^+ + e^- \rightleftarrows Au(s)$	1.69
Hydrogen	$2H^+ + 2e^- \rightleftarrows H_2(g)$	0 (by definition)
	$2D^+ + 2e^- \rightleftarrows D_2(g)$	−0.01
	$2H_2O + 2e^- \rightleftarrows H_2(g) + 2OH^-$	−0.83
Indium	$In^{3+} + 2e^- \rightleftarrows In^+$	−0.44
	$In^{3+} + 3e^- \rightleftarrows In(s)$	−0.34
Iodine	$2IO_3^- + 12H^+ + 10e^- \rightleftarrows I_2(s) + 6H_2O$	1.21
	$2HOI + 2H^+ + 2e^- \rightleftarrows I_2(s) + 2H_2O$	1.43
	$I_2(s) + 2e^- \rightleftarrows 2I^-$	0.54
	$I_2(aq) + 2e^- \rightleftarrows 2I^-$	0.62
	$I_3^- + 2e^- \rightleftarrows 3I^-$	0.54
Iron	$Fe^{3+} + e^- \rightleftarrows Fe^{2+}$	0.77
	$Fe(CN)_6^{3-} + e^- \rightleftarrows Fe(CN)_6^{4-}$	0.28
	$Fe(phen)_3^{3+} + e^- \rightleftarrows Fe(phen)_3^{2+}$ [b]	1.18
	$Fe^{3+} + 3e^- \rightleftarrows Fe(s)$	−0.04
	$Fe^{2+} + 2e^- \rightleftarrows Fe(s)$	−0.44
	$Fe(EDTA)^- + e^- \rightleftarrows Fe(EDTA)^{2-}$ [c]	0.12

[b] phen = 1,10-phenanthroline.
[c] $EDTA^{4-}$ = ethylenediaminetetraacetate ion.

Element	Reaction	$E^\ominus$/V
Lanthanum	$La^{3+} + 3e^- \rightleftarrows La(s)$	−2.38
Lead	$Pb^{4+} + 2e^- \rightleftarrows Pb^{2+}$	1.69
	$PbO_2(s) + 4H^+ + 2e^- \rightleftarrows Pb^{2+} + 2H_2O$	1.46
	$PbO_2(s) + 4H^+ + SO_4^{2-} + 2e^- \rightleftarrows PbSO_4(s) + 2H_2O$	1.69
	$Pb^{2+} + 2e^- \rightleftarrows Pb(s)$	−0.13
	$PbO(s) + H_2O + 2e^- \rightleftarrows Pb(s) + 2OH^-$	−0.58
	$PbSO_4(s) + 2e^- \rightleftarrows Pb(s) + SO_4^{2-}$	−0.36
Lithium	$Li^+ + e^- \rightleftarrows Li(s)$	−3.04
	$Li^+ + Hg(l) + e^- \rightleftarrows Li(Hg)(l)$	−2.00
Magnesium	$Mg^{2+} + 2e^- \rightleftarrows Mg(s)$	−2.36
Manganese	$MnO_4^- + e^- \rightleftarrows MnO_4^{2-}$	0.56
	$MnO_4^- + 2H_2O + 3e^- \rightleftarrows MnO_2(s) + 4OH^-$	0.59
	$MnO_4^- + 4H^+ + 3e^- \rightleftarrows MnO_2(s) + 2H_2O$	1.69
	$MnO_4^- + 8H^+ + 5e^- \rightleftarrows Mn^{2+} + 4H_2O$	1.51
	$MnO_2(s) + 4H^+ + 2e^- \rightleftarrows Mn^{2+} + 2H_2O$	1.23
	$Mn^{3+} + e^- \rightleftarrows Mn^{2+}$	1.56
	$Mn^{2+} + 2e^- \rightleftarrows Mn(s)$	−1.18
Mercury	$2Hg^{2+} + 2e^- \rightleftarrows Hg_2^{2+}$	0.91
	$Hg^{2+} + 2e^- \rightleftarrows Hg(l)$	0.85
	$Hg_2^{2+} + 2e^- \rightleftarrows 2Hg(l)$	0.80
	$Hg_2Cl_2(s) + 2e^- \rightleftarrows 2Hg(l) + 2Cl^-$	0.27[d]
	$Hg_2Br_2(s) + 2e^- \rightleftarrows 2Hg(l) + 2Br^-$	0.14
	$Hg_2I_2(s) + 2e^- \rightleftarrows 2Hg(l) + 2I^-$	−0.04
Nickel	$Ni^{2+} + 2e^- \rightleftarrows Ni(s)$	−0.24
	$Ni(NH_3)_6^{2+} + 2e^- \rightleftarrows Ni(s) + 6NH_3$	−0.48

[d] The potential of the 'saturated calomel' reference electrode KCl(saturated) | $Hg_2Cl_2(s)$, Hg relative to the standard hydrogen electrode is 0.2444 V at 298 K.

Element	Reaction	$E^\ominus$/V
Nitrogen	$2NO_3^- + 4H^+ + 2e^- \rightleftarrows N_2O_4(g) + 2H_2O$	0.80
	$NO_3^- + 3H^+ + 2e^- \rightleftarrows HNO_2 + H_2O$	0.94
	$NO_3^- + 4H^+ + 3e^- \rightleftarrows NO(g) + 2H_2O$	0.96
	$NO_3^- + 10H^+ + 8e^- \rightleftarrows NH_4^+ + 3H_2O$	0.88
	$HNO_2 + H^+ + e^- \rightleftarrows NO(g) + H_2O$	0.98
	$2NO + 2H^+ + 2e^- \rightleftarrows N_2O(g) + H_2O$	1.59
	$N_2O(g) + 2H^+ + 2e^- \rightleftarrows N_2(g) + H_2O$	1.77
	$3N_2(g) + 2H^+ + 2e^- \rightleftarrows 2HN_3$	−3.33
	$N_2(g) + 2H_2O + 4H^+ + 2e^- \rightleftarrows 2NH_3OH^+$	−1.87
	$N_2(g) + 5H^+ + 4e^- \rightleftarrows N_2H_5^+$	−0.21
	$NH_3OH^+ + 2H^+ + 2e^- \rightleftarrows NH_4^+ + H_2O$	1.35
	$N_2H_5^+ + 3H^+ + 2e^- \rightleftarrows 2NH_4^+$	1.25
Oxygen	$O_2(g) + 4H^+ + 4e^- \rightleftarrows 2H_2O$	1.23[e]
	$O_2(g) + 2H^+ + 2e^- \rightleftarrows H_2O_2$	0.70
	$O_2(g) + 2H_2O + 4e^- \rightleftarrows 4OH^-$	0.40[e]
	$O_3(g) + 2H^+ + 2e^- \rightleftarrows O_2(g) + H_2O$	2.07
	$H_2O_2 + 2H^+ + 2e^- \rightleftarrows 2H_2O$	1.76
Palladium	$Pd^{2+} + 2e^- \rightleftarrows Pd(s)$	0.92
Phosphorus	$H_3PO_4 + 2H^+ + 2e^- \rightleftarrows H_3PO_3 + H_2O$	−0.30
	$H_3PO_3 + 2H^+ + 2e^- \rightleftarrows H_3PO_2 + H_2O$	−0.48
	$H_3PO_2 + H^+ + e^- \rightleftarrows P(s, red) + 2H_2O$	−0.33
	$P(s, red) + 3H^+ + 3e^- \rightleftarrows PH_3(g)$	−0.09
	$P(s, white) + 3H^+ + 3e^- \rightleftarrows PH_3(g)$	−0.05

[e] The potential of the oxygen electrode at pH 7.00 is 0.82 V.

49 STANDARD ELECTRODE POTENTIALS *(continued)*

Element	Reaction	$E^{\ominus}$/V
Platinum	$PtCl_6^{2-} + 2e^- \rightleftharpoons PtCl_4^{2-} + 2Cl^-$	0.73
	$Pt^{2+} + 2e^- \rightleftharpoons Pt(s)$	1.18
	$PtCl_4^{2-} + 2e^- \rightleftharpoons Pt(s) + 4Cl^-$	0.76
Potassium	$K^+ + e^- \rightleftharpoons K(s)$	−2.94
Radium	$Ra^{2+} + 2e^- \rightleftharpoons Ra(s)$	−2.80
Rubidium	$Rb^+ + e^- \rightleftharpoons Rb(s)$	−2.94
Scandium	$Sc^{3+} + 3e^- \rightleftharpoons Sc(s)$	−2.09
Selenium	$SeO_4^{2-} + 4H^+ + 2e^- \rightleftharpoons H_2SeO_3 + H_2O$	1.15
	$H_2SeO_3 + 4H^+ + 4e^- \rightleftharpoons Se(s) + 3H_2O$	0.74
	$Se(s) + 2H^+ + 2e^- \rightleftharpoons H_2Se(g)$	−0.08
	$Se(s) + 2e^- \rightleftharpoons Se^{2-}$	−0.67
Silicon	$SiO_2(s) + 4H^+ + 4e^- \rightleftharpoons Si(s) + 2H_2O$	−0.99
	$Si(s) + 4H^+ + 4e^- \rightleftharpoons SiH_4(g)$	−0.15
Silver	$Ag_2O_3(s) + 6H^+ + 4e^- \rightleftharpoons 2Ag^+ + 3H_2O$	1.76
	$Ag^{2+} + e^- \rightleftharpoons Ag^+$	1.99
	$Ag_2O_2(s) + H_2O + 2e^- \rightleftharpoons Ag_2O(s) + 2OH^-$	0.60
	$Ag^+ + e^- \rightleftharpoons Ag(s)$	0.80
	$AgCN(s) + e^- \rightleftharpoons Ag(s) + CN^-$	−0.12
	$Ag(CN)_2^- + e^- \rightleftharpoons Ag(s) + 2CN^-$	−0.41
	$Ag(NH_3)_2^+ + e^- \rightleftharpoons Ag(s) + 2NH_3$	0.37
	$Ag_2O(s) + H_2O + 2e^- \rightleftharpoons 2Ag(s) + 2OH^-$	0.34
	$Ag_2S(s) + 2H^+ + 2e^- \rightleftharpoons 2Ag(s) + H_2S(g)$	−0.04
	$AgCl(s) + e^- \rightleftharpoons Ag(s) + Cl^-$	0.22
	$AgBr(s) + e^- \rightleftharpoons Ag(s) + Br^-$	0.07
	$AgI(s) + e^- \rightleftharpoons Ag(s) + I^-$	−0.15
Sodium	$Na^+ + e^- \rightleftharpoons Na(s)$	−2.71
	$Na^+ + Hg(l) + e^- \rightleftharpoons Na(Hg)(l)$	−1.84
Strontium	$Sr^{2+} + 2e^- \rightleftharpoons Sr(s)$	−2.90
Sulfur	$S_2O_8^{2-} + 2e^- \rightleftharpoons 2SO_4^{2-}$	2.01
	$SO_4^{2-} + 4H^+ + 2e^- \rightleftharpoons SO_2(aq) + 2H_2O$	0.16
	$2H_2SO_3 + 2H^+ + 4e^- \rightleftharpoons S_2O_3^{2-} + 3H_2O$	0.40
	$SO_2(aq) + 4H^+ + 4e^- \rightleftharpoons S(s) + 2H_2O$	0.45
	$S_4O_6^{2-} + 2e^- \rightleftharpoons 2S_2O_3^{2-}$	0.08
	$S(s) + 2e^- \rightleftharpoons S^{2-}$	−0.45
	$S(s) + 2H^+ + 2e^- \rightleftharpoons H_2S(g)$	0.17
Tellurium	$Te(OH)_6(s) + 2H^+ + 2e^- \rightleftharpoons TeO_2(s) + 4H_2O$	1.02
	$TeO_2(s) + 4H^+ + 4e^- \rightleftharpoons Te(s) + 2H_2O$	0.53
	$Te(s) + 2H^+ + 2e^- \rightleftharpoons H_2Te(g)$	−0.44
	$Te(s) + 2e^- \rightleftharpoons Te^{2-}$	−0.90
Thallium	$Tl^{3+} + 2e^- \rightleftharpoons Tl^+$	1.28
	$Tl^+ + e^- \rightleftharpoons Tl(s)$	−0.34
	$TlCl(s) + e^- \rightleftharpoons Tl(s) + Cl^-$	−0.56
	$TlBr(s) + e^- \rightleftharpoons Tl(s) + Br^-$	−0.66
	$TlI(s) + e^- \rightleftharpoons Tl(s) + I^-$	−0.77
Tin	$Sn^{4+} + 2e^- \rightleftharpoons Sn^{2+}$	0.15
	$Sn^{2+} + 2e^- \rightleftharpoons Sn(s)$	−0.14
Titanium	$TiOH^{3+} + H^+ + e^- \rightleftharpoons Ti^{3+} + H_2O$	−0.06
	$Ti^{3+} + 3e^- \rightleftharpoons Ti(s)$	−1.37
	$Ti^{2+} + 2e^- \rightleftharpoons Ti(s)$	−1.60

Element	Reaction	$E^{\ominus}$/V
Uranium	$UO_2^{2+} + e^- \rightleftarrows UO_2^+$	0.16
	$UO_2^+ + 4H^+ + e^- \rightleftarrows U^{4+} + 2H_2O$	0.39
	$U^{4+} + e^- \rightleftarrows U^{3+}$	−0.58
	$U^{3+} + 3e^- \rightleftarrows U(s)$	−1.64
Vanadium	$VO_2^+ + 2H^+ + e^- \rightleftarrows VO^{2+} + H_2O$	1.00
	$VO^{2+} + 2H^+ + e^- \rightleftarrows V^{3+} + H_2O$	0.34
	$V^{3+} + e^- \rightleftarrows V^{2+}$	−0.26
	$V^{2+} + 2e^- \rightleftarrows V(s)$	−1.13
Zinc	$Zn^{2+} + 2e^- \rightleftarrows Zn(s)$	−0.76
	$Zn(NH_3)_4^{2+} + 2e^- \rightleftarrows Zn(s) + 4NH_3$	−1.03
	$Zn(OH)_4^{2-} + 2e^- \rightleftarrows Zn(s) + 4OH^-$	−1.20

50 POTENTIALS OF COMMON REFERENCE ELECTRODES AT 298 K

Electrode	Potential versus standard hydrogen electrode /V
Calomel **$KCl(aq)|Hg_2Cl_2(s)|Hg(l)$**	
KCl(sat) (SCE)	0.2444
3.5 M KCl	0.250
1.0 M KCl	0.2824
0.1 M KCl	0.3358
Silver–silver chloride **$KCl(aq)|AgCl(s)|Ag(s)$**	
KCl(sat)	0.1989
3.5 M KCl	0.205
3.0 M KCl	0.210
1.0 M KCl	0.2272
0.1 M KCl	0.2901
Silver–silver sulfate **$H_2SO_4(aq)|Ag_2SO_4(s)|Ag(s)$**	
K_2SO_4(sat)	0.69
1.0 M H_2SO_4	0.71
0.5 M H_2SO_4	0.72
Mercury–mercuric oxide **$NaOH(aq)|HgO(s)|Hg(l)$**	
0.1 M NaOH	0.165
1.0 M NaOH	0.140
Mercury–mercurous sulfate **$H_2SO_4(aq)|Hg_2SO_4(s)|Hg(l)$**	
0.5 M H_2SO_4	0.682
1.0 M H_2SO_4	0.674
K_2SO_4(sat)	0.65

51 CONVERSION FACTORS (mV) BETWEEN COMMON REFERENCE ELECTRODES IN ACETONITRILE

All data refer to 298 K.

conversion ↓from\to→	Ag^+ (0.1 M $AgNO_3$[a])/Ag	Ag^+ (0.01 M $AgNO_3$[a])/Ag	Ag^+ (0.001 M $AgNO_3$[a])/Ag	Fc^+/Fc[b]	NHE[c]	SCE[d]	SHE[e]
Ag^+ (0.1 M $AgNO_3$[a])/Ag	0	+45	+97	−37	+593	+343	+587
Ag^+ (0.01 M $AgNO_3$[a])/Ag	−45	0	+52	−87	+548	+298	+542
Ag^+ (0.001 M $AgNO_3$[a])/Ag	−97	−52	0	−133	+496	+246	+490
Fc^+/Fc[b]	+37	+87	+133	0	+630	+380	+624
NHE[c]	−593	−548	−496	−630	0	−250	−6
SCE[d]	−343	−298	−246	−380	+250	0	+244
SHE[e]	−587	−542	−490	−624	+6	−244	0

[a] 0.1 M tetraethylammonium perchlorate (TEAP), CH_3CN
[b] Fc^+/Fc ferrocenium/ferrocene couple
[c] Normal hydrogen electrode
[d] Saturated calomel electrode
[e] Standard hydrogen electrode

52 GLOBALLY HARMONISED SYSTEM OF CLASSIFICATION AND LABELLING OF CHEMICALS (GHS)

Previous editions of *SI Chemical Data* have included a comprehensive chemical hazard classification system. This edition presents the new 'Globally Harmonised System of Classification and Labelling of Chemicals', or GHS. The GHS is a single, internationally agreed system of chemical classification and hazard communication through the use of labelling and safety data sheets. The GHS is published by the United Nations and is sometimes referred to as 'the purple book'. It includes harmonised criteria for the classification of physical hazards, health hazards and environmental hazards.

International implementation of the GHS is a staggered process and some countries may be in a transition phase or not have begun introducing the new system. Therefore, the GHS information presented herein may not be completely relevant to all chemical users. Readers are encouraged to search for chemical safety information relevant to their particular legislative requirements. Information about the extent of adoption of the GHS is available at www.unece.org/trans/danger/publi/ghs/implementation_e.html.

The GHS addresses classification of chemicals by types of hazard and includes harmonised hazard communication elements, including labels and safety data sheets. The classification system aims to ensure that information on physical hazards and toxicity from chemicals is available in order to enhance the protection of human health and the environment during the handling, transport and use of these chemicals. In addition, the GHS provides a basis for harmonisation of rules and regulations on chemicals at the national, regional and worldwide levels, an important factor also for trade facilitation. Further information is available at www.unece.org/trans/danger/publi/ghs/ghs_welcome_e.html and www.sigmaaldrich.com/safety-center/globally-harmonized.html.

The GHS is built on 16 physical, 10 health and 3 environmental hazard classes and includes four communication elements. 'Hazard communication' is a term used to describe how critical information about the hazards of chemicals and any precautions necessary to ensure safe storage, handling and disposal are conveyed to users of chemicals. In the GHS, hazards are communicated to chemical users through a combination of symbols and words in the form of signal words, hazard statements and precautionary statements. These are intended to appear on labels and in material safety data sheets (MSDS). Hazard statements are assigned to a class and category that describes the nature of the hazards of a chemical, including, where appropriate, the degree of hazard. The communication elements in the GHS are:

- nine *hazard pictograms*, which represent the physical, health and environmental hazards (see below). *The GHS also recognises the equivalent pictograms that apply to dangerous goods and local codes in various countries.*
- two *signal words*, 'Danger' and 'Warning', to indicate the relative severity of a hazard. 'Danger' is used for the more severe or significant hazards, while 'Warning' is used for less severe hazards.
- 72 individual and 17 combined *hazard statements*. These are assigned a unique code, which consists of one letter and three digits as follows:
  (a) the letter 'H' (for 'hazard statement')
  (b) a number designating the type of hazard:
    – '2' for physical hazards
    – '3' for health hazards
    – '4' for environmental hazards
  (c) a further two digits, corresponding to the sequential numbering of hazards arising from the intrinsic properties of the substance or mixture.

  For example, explosive properties are given codes from H200 to H210, flammability properties are given codes from H220 to H230.
- 116 individual and 33 combined *precautionary statements*. These are assigned a unique code, which consists of one letter and three digits as follows:
  (a) the letter 'P' (for 'precautionary statement')
  (b) a number designating the type of precautionary statement:
    – '1' for general precautionary statements
    – '2' for prevention precautionary statements
    – '3' for response precautionary statements
    – '4' for storage precautionary statements
    – '5' for disposal precautionary statements
  (c) a further two digits, corresponding to the sequential numbering of precautionary statements.

  Precautionary statements describe the recommended measures that should be taken to minimise or prevent adverse effects resulting from exposure or improper storage or handling of a hazardous chemical. The GHS categorises precautionary statements according to whether they relate to prevention, response, storage or disposal.

Pictograms

The nine hazard pictograms in the GHS representing the physical, health and environmental hazards are shown in the table below.

Description	Pictogram	Hazard class and hazard category
Exploding bomb		• Unstable explosives • Explosives of Divisions 1.1, 1.2, 1.3, 1.4 • Self reactive substances and mixtures, Types A, B • Organic peroxides, Types A, B
Flame		• Flammable gases, category 1 • Flammable aerosols, categories 1, 2 • Flammable liquids, categories 1, 2, 3 • Flammable solids, categories 1, 2 • Self-reactive substances and mixtures, Types B, C, D, E, F • Pyrophoric liquids, category 1 • Pyrophoric solids, category 1 • Self-heating substances and mixtures, categories 1, 2 • Substances and mixtures, which in contact with water, emit flammable gases, categories 1, 2, 3 • Organic peroxides, Types B, C, D, E, F
Flame over circle		• Oxidising gases, category 1 • Oxidising liquids, categories 1, 2, 3
Gas cylinder		• Gases under pressure: – Compressed gases – Liquefied gases – Refrigerated liquefied gases – Dissolved gases
Corrosion		• Corrosive to metals, category 1 • Skin corrosion, categories 1A, 1B, 1C • Serious eye damage, category 1
Skull and crossbones		• Acute toxicity (oral, dermal, inhalation), categories 1, 2, 3
Exclamation mark		• Acute toxicity (oral, dermal, inhalation), category 4 • Skin irritation, category 2 • Eye irritation, category 2 • Skin sensitisation, category 1 • Specific target organ toxicity — Single exposure, category 3
Health hazard		• Respiratory sensitisation, category 1 • Germ cell mutagenicity, categories 1A, 1B, 2 • Carcinogenicity, categories 1A, 1B, 2 • Reproductive toxicity, categories 1A, 1B, 2 • Specific target organ toxicity — Single exposure, categories 1, 2 • Specific target organ toxicity — Repeated exposure, categories 1, 2 • Aspiration Hazard, category 1
Environment		• Hazardous to the aquatic environment – Acute hazard, category 1 – Chronic hazard, categories 1, 2

Source: © Sigma-Aldrich Co. LLC. Used with permission.

The information in this section draws on information published by:

- Sigma Aldrich, www.sigmaaldrich.com/safety-center/globally-harmonized.html. © Sigma-Aldrich Co. LLC. Used with permission. Users of chemicals listed in this book or elsewhere should access this site or another site with MSDSs relevant to the chemicals being used. Users should consult MSDS sites specific to their particular country and legislative requirements.
- Safe Work Australia, www.safeworkaustralia.gov.au/sites/swa/whs-information/hazardous-chemicals/ghs/pages/ghs.

Other useful sites for information include:

- www.unece.org/trans/danger/publi/ghs/ghs_rev03/03files_e.html
- www.unece.org/trans/danger/publi/ghs/ghs_welcome_e.html
- www2.unitar.org/cwm/publications/cw/ghs/GHS_Companion_Guide_final_June2010.pdf.

This appendix on the Globally Harmonised System of Classification and Labelling of Chemicals should be read in conjunction with the MSDS information included in the next appendix, 53.

53 INTERPRETATION OF A MATERIAL SAFETY DATA SHEET (MSDS)

A material safety data sheet (MSDS) provides basic information on the properties and potential hazards of a chemical or a material, indications on how to use it safely, and steps to take if there is an emergency involving that substance. An MSDS is often specific to the country in which it is used; readers are therefore encouraged to familiarise themselves with the format and legislative requirements for these data in their particular country.

The following outline presents the different sections of a typical MSDS with an indication of the information contained in those sections. Not every MSDS will have an identical layout; however, they will contain some, if not all, of the following information.

MATERIAL SAFETY DATA SHEET

1 **Product and company identification**
 - **(i)** Product identifiers:
   - Product name
   - Chemical class
   - Product number
 - **(ii)** Name, address and emergency telephone number of the supplier of the MSDS
 - **(iii)** Name, address and emergency telephone number of the manufacturer of the material
 - **(iv)** Product use
 - **(v)** Date of preparation of the MSDS

2 **Hazards identification**
 - **(i)** Globally Harmonised System (GHS) classification
 - **(ii)** GHS label elements including the pictogram and the precautionary and hazard statements
 - **(iii)** An indication of other hazards
 - **(iv)** Potential health effects
   - Relevant routes of exposure
   - Adverse health effects from exposure to product or materials of product
   - Length of exposure
   - Severity of effect
   - Target organ
   - Type of effect
 - **(v)** Signs and symptoms

3 Composition/information on materials

- **(i)** Synonyms
- **(ii)** Formula
- **(iii)** Chemical Abstracts Service (CAS) registry number of hazardous materials (present at 1.0%, or 0.1% as appropriate, by weight). Each CAS registry number is unique, designating only one substance. The number has no chemical significance, but links to a wealth of information about a specific chemical substance.
- **(iv)** European Commission (EC) number. The EC number is a seven-digit number assigned to chemical substances for regulatory purposes within the European Union by the regulatory authorities.
- **(v)** Materials with unknown toxicological properties
- **(vi)** Generic chemical identity, registry number, and date of claim for trade secret materials (if a legislative requirement for the particular country)

4 First aid measures

- **(i)** A description of the appropriate first aid measures following exposure:
  - inhaled
  - skin contact
  - eye contact
  - swallowed
- **(ii)** Immediate symptoms
- **(iii)** Note to physician
- **(iv)** Medical conditions likely to be aggravated by exposure

5 Fire fighting measures

- **(i)** Extinguishing media and fire fighting procedures
- **(ii)** Flash point and method
- **(iii)** Upper and lower flammable (explosive) limits in air
- **(iv)** Autoignition temperature
- **(v)** Hazardous combustion products
- **(vi)** Conditions under which flammability could occur
- **(vii)** Sensitivity to explosion by mechanical impact
- **(viii)** Sensitivity to explosion by static discharge

6 Accidental release measures

- **(i)** Procedures for dealing with a release or spill

7 Handling and storage

- **(i)** Precautions for safe handling
- **(ii)** Conditions for safe storage, including information on incompatible storage

8 Exposure controls/personal protection

- **(i)** Control parameters — the LD50 and/or LC50 for materials. (LD stands for 'lethal dose'. LD50 is the amount of a chemical or material, given in one dose, which causes the death of 50% of a group of test specimens. LC stands for 'lethal concentration'. LC50 data usually refer to the concentration of a chemical in air (but can refer to water). LC50 is the concentration of the chemical in air that kills 50% of the test specimens in a given time.)
- **(ii)** Exposure controls
- **(iii)** Personal protective equipment
  - eye
  - skin
  - body
  - respiratory

9 Physical and chemical properties

- **(i)** Appearance
- **(ii)** Physical state
- **(iii)** Odour
- **(iv)** Odour threshold
- **(v)** pH
- **(vi)** Boiling point
- **(vii)** Freezing/melting point
- **(viii)** Evaporation rate
- **(ix)** Vapour pressure and reference temperature
- **(x)** Vapour density
- **(xi)** Specific gravity or density
- **(xii)** Partition coefficient
- **(xiii)** Solubility in water
- **(xiv)** Viscosity

10 Stability and reactivity

- **(i)** Chemical stability
- **(ii)** Reactivity
- **(iii)** Incompatibility with other materials
- **(iv)** Hazardous decomposition products
- **(v)** Hazardous polymerisation

11 Toxicological information

- **(i)** Effects of short-term exposure
- **(ii)** Effects of long-term exposure
- **(iii)** Irritancy
- **(iv)** Sensitisation
- **(v)** Carcinogenicity

(vi) Reproductive toxicity
(vii) Teratogenicity (relating to substances or agents that can interfere with normal embryonic development)
(viii) Mutagenicity (relating to the property of being able to induce genetic mutation)
(ix) Name of toxicologically synergistic products

12 Ecological information
(i) Toxicity
(ii) Persistence and degradability
(iii) Bioaccumulation potential
(iv) Mobility in soil
(v) Other adverse effects

13 Disposal considerations
(i) Waste treatment and disposal information

14 Transport information
(i) UN number (four-digit numbers that identify hazardous substances (explosives, flammable liquids, toxic substances, etc.)) in the framework of international transport
(ii) UN proper shipping name
(iii) Transport hazard class
(iv) Packaging group
(v) Environmental hazards
(vi) Precautions for user

15 Regulatory information
(i) Safety, health and environmental regulations etc. specific for the material
As mentioned in the introduction, an MSDS is often specific to the country in which it is used. Different countries will have different legislative requirements regarding the information contained in an MSDS. Familiarisation with the legislative requirements and format of the MSDS is encouraged.

16 Other information

Example MSDS

The following is an MSDS for ethanol.

SIGMA-ALDRICH

sigma-aldrich.com

SAFETY DATA SHEET

Version 4.5

Revision Date 04.02.2013

Print Date 14.04.2013

1. IDENTIFICATION OF THE SUBSTANCE/MIXTURE AND OF THE COMPANY/UNDERTAKING

1.1 Product identifiers

Product name : Ethanol

Product Number : E7023
Brand : Sigma-Aldrich

1.2 Other means of identification

Ethyl alcohol

1.3 Relevant identified uses of the substance or mixture and uses advised against

Identified uses : Laboratory chemicals, Manufacture of substances

1.4 Details of the supplier of the safety data sheet

Company : Sigma-Aldrich Pty. Ltd.
12 Anella Avenue
CASTLE HILL NSW 2154
AUSTRALIA

Telephone : +61 2 9841 0555 (1800 800 097)
Fax : +61 2 9841 0500 (1800 800 096)

1.5 Emergency telephone number

Emergency Phone # : +44 (0)8701 906777 (1800 448 465)

2. HAZARDS IDENTIFICATION

2.1 GHS Classification

Flammable liquids (Category 2)
Specific target organ toxicity – single exposure (Category 3)

2.2 GHS Label elements, including precautionary statements

Pictogram

Signal word: Danger

Hazard statement(s)

H225 Highly flammable liquid and vapour.

H335 May cause respiratory irritation.

Precautionary statement(s)

Prevention

P210 Keep away from heat/sparks/open flames/hot surfaces. No smoking.

P233 Keep container tightly closed.

P240 Ground/bond container and receiving equipment.

P241 Use explosion-proof electrical/ventilating/lighting/equipment.

P242 Use only non-sparking tools.

P243 Take precautionary measures against static discharge.

P261 Avoid breathing dust/fume/gas/mist/vapours/spray.

P271 Use only outdoors or in a well-ventilated area.

P280 Wear protective gloves/protective clothing/eye protection/face protection.

Response

P303 + P361 + P353 IF ON SKIN (or hair): Remove/Take off immediately all contaminated clothing. Rinse skin with water/shower.

P304 + P340 IF INHALED: Remove victim to fresh air and keep at rest in a position comfortable for breathing.

P312 Call a POISON CENTER or doctor/physician if you feel unwell.

P370 + P378 In case of fire: Use dry sand, dry chemical or alcohol-resistant foam for extinction.

Storage

P403 + P233 Store in a well-ventilated place. Keep container tightly closed.

P403 + P235 Store in a well-ventilated place. Keep cool.

P405 Store locked up.

Disposal

P501 Dispose of contents/container to an approved waste disposal plant.

2.3 Other hazards – none

3. COMPOSITION/INFORMATION ON INGREDIENTS

3.1 Substances

Synonyms : Ethyl alcohol

Formula : C_2H_6O

Molecular Weight : 46.07 g/mol

CAS-No. : 64-17-5

EC-No. : 200-578-6

Index-No. : 603-002-00-5

Component	Classification	Concentration
Ethanol		
	Flam. Liq. 2; H225	-

For the full text of the H-Statements mentioned in this section, see section 16.

4. FIRST AID MEASURES

4.1 Description of first aid measures

General advice

Consult a physician. Show this safety data sheet to the doctor in attendance.

If inhaled

If breathed in, move person into fresh air. If not breathing, give artificial respiration. Consult a physician.

In case of skin contact
Wash off with soap and plenty of water. Consult a physician.

In case of eye contact
Rinse thoroughly with plenty of water for at least 15 minutes and consult a physician.

If swallowed
Do NOT induce vomiting. Never give anything by mouth to an unconscious person. Rinse mouth with water. Consult a physician.

4.2 Most important symptoms and effects, both acute and delayed
The most important known symptoms and effects are described in the labelling (see section 2.2) and/or in section 11.

4.3 Indication of any immediate medical attention and special treatment needed
no data available

5. FIREFIGHTING MEASURES

5.1 Extinguishing media
Suitable extinguishing media
Use water spray, alcohol-resistant foam, dry chemical or carbon dioxide.

5.2 Special hazards arising from the substance or mixture
no data available

5.3 Advice for firefighters
Wear self contained breathing apparatus for fire fighting if necessary.

5.4 Further information
Use water spray to cool unopened containers.

6. ACCIDENTAL RELEASE MEASURES

6.1 Personal precautions, protective equipment and emergency procedures
Use personal protective equipment. Avoid breathing vapours, mist or gas. Ensure adequate ventilation.
Remove all sources of ignition. Evacuate personnel to safe areas.
Beware of vapours accumulating to form explosive concentrations.
Vapours can accumulate in low areas.
For personal protection see section 8.

6.2 Environmental precautions
Prevent further leakage or spillage if safe to do so. Do not let product enter drains.

6.3 Methods and materials for containment and cleaning up
Contain spillage, and then collect with an electrically protected vacuum cleaner or by wet-brushing and place in container for disposal according to local regulations (see section 13).

6.4 Reference to other sections
For disposal see section 13.

7. HANDLING AND STORAGE

7.1 Precautions for safe handling
Avoid contact with skin and eyes. Avoid inhalation of vapour or mist.
Keep away from sources of ignition – No smoking. Take measures to prevent the build up of electrostatic charge.
For precautions see section 2.2.

7.2 Conditions for safe storage, including any incompatibilities
Store in cool place. Keep container tightly closed in a dry and well-ventilated place. Containers which are opened must be carefully resealed and kept upright to prevent leakage.
Hygroscopic.

7.3 Specific end use(s)
Apart from the uses mentioned in section 1.2 no other specific uses are stipulated.

8. EXPOSURE CONTROLS/PERSONAL PROTECTION

8.1 Control parameters

Occupational Exposure Limits

Component	CAS-No.	Value	Control parameters	Basis
Ethanol	64-17-5	TWA	1,000 ppm 1,880 mg/m3	Australia. Workplace Exposure Standards for Airborne Contaminants.
	Remarks	ACGIH is the documentation source		

8.2 Exposure controls

Appropriate engineering controls
Handle in accordance with good industrial hygiene and safety practice. Wash hands before breaks and at the end of workday.

Personal protective equipment

Eye/face protection

Face shield and safety glasses Use equipment for eye protection tested and approved under appropriate government standards such as NIOSH (US) or EN 166(EU).

Skin protection

Handle with gloves. Gloves must be inspected prior to use. Use proper glove removal technique (without touching glove's outer surface) to avoid skin contact with this product. Dispose of contaminated gloves after use in accordance with applicable laws and good laboratory practices. Wash and dry hands.

The selected protective gloves have to satisfy the specifications of EU Directive 89/686/EEC and the standard EN 374 derived from it.

Full contact
Material: butyl-rubber
Minimum layer thickness: 0.3 mm
Break through time: 480 min
Material tested: Butoject® (KCL 897 /Aldrich Z677647, Size M)

Splash contact
Material: Nitrile rubber
Minimum layer thickness: 0.2 mm
Break through time: 38 min
Material tested: Dermatril® P (KCL 743 /Aldrich Z677388, Size M)

Data source: KCL GmbH, D-36124 Eichenzell, phone +49 (0)6659 87300, e-mail sales@kcl.de, test method: EN374. If used in solution, or mixed with other substances, and under conditions which differ from EN 374, contact the supplier of the CE approved gloves. This recommendation is advisory only and must be evaluated by an industrial hygienist and safety officer familiar with the specific situation of anticipated use by our customers. It should not be construed as offering an approval for any specific use scenario.

Body protection

Impervious clothing. Flame retardant antistatic protective clothing. The type of protective equipment must be selected according to the concentration and amount of the dangerous substance at the specific workplace.

Respiratory protection

Where risk assessment shows air-purifying respirators are appropriate use a full-face respirator with multi-purpose combination (US) or type ABEK (EN 14387) respirator cartridges as a backup to engineering controls. If the respirator is the sole means of protection, use a full-face supplied air respirator. Use respirators and components tested and approved under appropriate government standards such as NIOSH (US) or CEN (EU).

9. PHYSICAL AND CHEMICAL PROPERTIES

9.1 Information on basic physical and chemical properties

a)	Appearance	Form: liquid, clear. Colour: colourless
b)	Odour	no data available
c)	Odour Threshold	no data available
d)	pH	no data available
e)	Melting point/freezing point	−144.0 °C
f)	Initial boiling point and boiling range	78.0 – 80.0 °C
g)	Flash point	14.0 °C – closed cup
h)	Evaporation rate	no data available
i)	Flammability (solid, gas)	no data available
j)	Upper/lower flammability or explosive limits	Upper explosion limit: 19%(V) Lower explosion limit: 3.3%(V)
k)	Vapour pressure	59.5 hPa at 20.0 °C
l)	Vapour density	no data available
m)	Relative density	0.7974 g/cm^3
n)	Water solubility	completely soluble
o)	Partition coefficient: n-octanol/water	no data available
p)	Auto-ignition temperature	363.0 °C
q)	Decomposition temperature	no data available

r) Viscosity — no data available
s) Explosive properties — no data available
t) Oxidizing properties — no data available

9.2 Other safety information
no data available

10. STABILITY AND REACTIVITY

10.1 Reactivity
no data available

10.2 Chemical stability
Stable under recommended storage conditions.

10.3 Possibility of hazardous reactions
no data available

10.4 Conditions to avoid
Heat, flames and sparks. Extremes of temperature and direct sunlight.

10.5 Incompatible materials
Alkali metals, Ammonia, Oxidizing agents, Peroxides

10.6 Hazardous decomposition products
Other decomposition products – no data available
In the event of fire: see section 5

11. TOXICOLOGICAL INFORMATION

11.1 Information on toxicological effects

Acute toxicity
LD50 Oral – rat – 7,060 mg/kg
Remarks: Lungs, Thorax, or Respiration: Other changes.

LC50 Inhalation – rat – 10 h – 20000 ppm

Skin corrosion/irritation
Skin – rabbit
Result: Irritating to skin – 24 h

Serious eye damage/eye irritation
Eyes – rabbit
Result: Mild eye irritation – 24 h
(Draize Test)

Respiratory or skin sensitisation
no data available

Germ cell mutagenicity
no data available

Carcinogenicity

Carcinogenicity – mouse – Oral
Tumorigenic: Equivocal tumorigenic agent by RTECS criteria.
Liver: Tumors. Blood: Lymphomas including Hodgkin's disease.
IARC: No component of this product present at levels greater than or equal to 0.1% is identified as probable, possible or confirmed human carcinogen by IARC.

Reproductive toxicity
Reproductive toxicity – Human – female – Oral
Effects on Newborn: Apgar score (human only). Effects on Newborn: Other neonatal measures or effects.
Effects on Newborn: Drug dependence.

Specific target organ toxicity – single exposure
no data available

Specific target organ toxicity – repeated exposure
no data available

Aspiration hazard
no data available

Additional information
RTECS: KQ6300000

Central nervous system depression, narcosis, Damage to the heart. To the best of our knowledge, the chemical, physical, and toxicological properties have not been thoroughly investigated.

Heart – Irregularities – Based on Human Evidence

12. ECOLOGICAL INFORMATION

12.1 Toxicity
no data available

12.2 Persistence and degradability
no data available

12.3 Bioaccumulative potential
no data available

12.4 Mobility in soil
no data available

12.5 Results of PBT and vPvB assessment
PBT/vPvB assessment not available as chemical safety assessment not required/not conducted

12.6 Other adverse effects
no data available

13. DISPOSAL CONSIDERATIONS

13.1 Waste treatment methods

Product
Burn in a chemical incinerator equipped with an afterburner and scrubber but exert extra care in igniting as this material is highly flammable. Offer surplus and non-recyclable solutions to a licensed disposal company.

Contaminated packaging
Dispose of as unused product.

14. TRANSPORT INFORMATION

14.1 UN number
ADR/RID: 1170 IMDG: 1170 IATA-DGR: 1170

14.2 UN proper shipping name
ADR/RID : ETHANOL
IMDG : ETHANOL
IATA-DGR : Ethanol

14.3 Transport hazard class(es)
ADR/RID: 3 IMDG: 3 IATA-DGR: 3

14.4 Packaging group
ADR/RID: II IMDG: II IATA-DGR: II

14.5 Environmental hazards
ADR/RID: no IMDG Marine pollutant: no IATA-DGR: no

14.6 Special precautions for user
no data available

15. REGULATORY INFORMATION

15.1 Safety, health and environmental regulations/legislation specific for the substance or mixture

Standard for the Uniform Scheduling of Medicines and Poisons
no data available

Carcinogen classification under WHS Regulation 2011, Schedule 10
Not listed

Notification status

AICS: On the inventory, or in compliance with the inventory
DSL: All components of this product are on the Canadian DSL.
ENCS: On the inventory, or in compliance with the inventory
IECSC: On the inventory, or in compliance with the inventory
ISHL: Not in compliance with the inventory – Ethanol
KECI: On the inventory, or in compliance with the inventory
NZIoC: On the inventory, or in compliance with the inventory
PICCS: On the inventory, or in compliance with the inventory

16. OTHER INFORMATION

Full text of H-Statements referred to under sections 2 and 3.

Flam. Liq. Flammable liquids
H225 Highly flammable liquid and vapour.

Further information
Copyright 2013 Sigma-Aldrich Co. LLC. License granted to make unlimited paper copies for internal use only.
The above information is believed to be correct but does not purport to be all inclusive and shall be used only as a guide. The information in this document is based on the present state of our knowledge and is applicable to the product with regard to appropriate safety precautions. It does not represent any guarantee of the properties of the product. Sigma-Aldrich Corporation and its Affiliates shall not be held liable for any damage resulting from handling or from contact with the above product. See www.sigmaaldrich.com and/or the reverse side of invoice or packing slip for additional terms and conditions of sale.

Source: © Sigma-Aldrich Co. LLC. Used with permission.

54 COMMON FORMULAE AND DEFINITIONS

acidity constant for the acid–base reaction:

$$\mathbf{HA(aq) + H_2O(l) \rightleftharpoons H_3O^+(aq) + A^-(aq)}$$

$$K_a = \frac{[H_3O^+][A^-]}{[HA]}$$

$$pK_a = -\log K_a$$

Arrhenius equation

$$k = Ae^{-E_a/RT}$$

$$\ln\frac{k_2}{k_1} = -\frac{E_a}{R}\left[\frac{1}{T_2} - \frac{1}{T_1}\right]$$

basicity constant for the acid–base reaction:

$$\mathbf{B(aq) + H_2O(l) \rightleftharpoons BH^+(aq) + OH^-(aq)}$$

$$K_b = \frac{[BH^+][OH^-]}{[B]}$$

$$pK_b = -\log K_b$$

boiling point elevation

$$\Delta T_b = K_b b$$

cell potential

$$E_{cell} = E_R - E_L$$

$$E^\ominus_{cell} = E^\ominus_R - E^\ominus_L$$

$$E^\ominus_{cell} = \frac{RT}{zF}\ln K_c$$

concentration molality

$$b = \frac{\text{amount of solute (mol)}}{\text{mass of solvent (kg)}}$$

molarity

$$c = \frac{n}{V}$$

cumulative formation constant for the formation reaction:

$$\mathbf{M^{x+}(aq) + \mathit{n}L(aq) \rightleftharpoons [ML_\mathit{n}]^{x+}(aq)}$$

$$\beta_n = \frac{[ML_n{}^{x+}(aq)]}{[M^{x+}(aq)][L(aq)]^n}$$

energy of a photon

$$E = h\nu = \frac{hc}{\lambda}$$

enthalpy

$$H = U + pV$$

entropy of reaction, standard for the general reaction:

$$\boldsymbol{aA + bB \rightarrow cC + dD}$$

$$\Delta_r S^\ominus = cS^\ominus_C + dS^\ominus_D - (aS^\ominus_A + bS^\ominus_B)$$

entropy and enthalpy, relationship

$$\Delta S_{surroundings} = \frac{-\Delta H_{system}}{T}$$

equilibrium constant expression for the solution-phase reaction:

$$\boldsymbol{aA + bB \rightleftharpoons cC + dD}$$

$$K_c = \frac{\left(\frac{[C]_e}{c^\ominus}\right)^c\left(\frac{[D]_e}{c^\ominus}\right)^d}{\left(\frac{[A]_e}{c^\ominus}\right)^a\left(\frac{[B]_e}{c^\ominus}\right)^b}$$

$$K_c = \frac{[C]^c[D]^d}{[A]^a[B]^b}, \quad \text{when } c^\ominus = 1\text{ mol L}^{-1}$$

equilibrium constant for the gas-phase reaction:

$$\boldsymbol{aA + bB \rightleftharpoons cC + dD}$$

$$K_p = \frac{\left(\frac{p_C}{p^\ominus}\right)^c\left(\frac{p_D}{p^\ominus}\right)^d}{\left(\frac{p_A}{p^\ominus}\right)^a\left(\frac{p_B}{p^\ominus}\right)^b}$$

Faraday's law

$$\text{amount of product} = \frac{It}{zF}$$

first law of thermodynamics

$$\Delta U = q + w$$

freezing point depression

$\Delta T_f = K_f b$

gas density

$$\rho_{gas} = \frac{m}{V} = \frac{pM}{RT}$$

Gibbs energy

$G = H - TS$

$\Delta G = \Delta H - T\Delta S$

$\Delta G = -zFE_{cell}$

$\Delta G^{\ominus} = -zFE^{\ominus}_{cell}$

$\Delta_r G^{\ominus} = -RT\ln k$

$\Delta_r G = \Delta_r G^{\ominus} + RT\ln Q$

change in standard Gibbs energy

$\Delta G^{\ominus} = \Delta H^{\ominus} - T\Delta S^{\ominus}$

for: $aA + bB \rightarrow cC + dD$

$\Delta_r G^{\ominus} = c\Delta_f G^{\ominus}{}_C + d\Delta_f G^{\ominus}{}_D - (a\Delta_f G^{\ominus}{}_A + b\Delta_f G^{\ominus}{}_B)$

half-life for a first-order reaction

$$t_{\frac{1}{2}} = \frac{\ln 2}{k} = \frac{0.693}{k}$$

half-life for a second-order reaction

$$t_{\frac{1}{2}} = \frac{1}{k[A]_0}$$

half-life for a zero-order reaction

$$t_{\frac{1}{2}} = \frac{[A]_0}{2k}$$

heat

$q = C\Delta T$

$q = cm\Delta T$

heat of reaction

$\Delta H = q_p$ (constant pressure)

$\Delta_r U = q_v$ (constant volume)

Henderson–Hasselbalch equation

$$\text{pH} = \text{p}K_a + \log\frac{[\text{A}^-]}{[\text{HA}]}$$

Henry's law

$c_{gas} = k_H p_{gas}$(constant T)

$$\frac{c_1}{p_1} = \frac{c_2}{p_2}$$

Hess's law equation for the general reaction:

aA* + *bB* → *cC* + *dD

$\Delta_r H^{\ominus} = c\Delta_f H^{\ominus}{}_C + d\Delta_f H^{\ominus}{}_D - (a\Delta_f H^{\ominus}{}_A + b\Delta_f H^{\ominus}{}_B)$

ideal gas equation

$pV = nRT$

$p_i V_i = p_f V_f$ (constant n and T)

index of hydrogen deficiency (IHD)

$$\text{IHD} = \frac{(\text{H}_{reference} - \text{H}_{molecule})}{2}$$

integrated rate law for a first-order reaction

$\ln [A]_t = -kt + \ln[A]_0$

$[A]_t = [A]_0 e^{-kt}$

integrated rate law for a second-order reaction

$$\frac{1}{[A]_t} = kt + \frac{1}{[A]_0}$$

integrated rate law for a zero-order reaction

$[A]_t = -kt + [A]_0$

isoelectric point of an amino acid

$\text{pI} = \frac{1}{2}(\text{p}K_a \text{ of } \alpha\text{–COOH} + \text{p}K_a \text{ of } \alpha\text{–NH}_3{}^+)$

kinetic energy

$E_{kinetic} = \frac{1}{2}mu^2$

$E_{kinetic,\ molar} = \frac{3}{2}RT$, for 1 mole of gas molecules

$\overline{E}_{kinetic} = \frac{3RT}{2N_A}$, average of gas molecules

K_p and K_c, relationship

$$K_p = K_c\left(\frac{RT}{p^{\ominus}}\right)^{\Delta n_g}$$

law of radioactive decay

$$\text{activity} = -\frac{\Delta N}{\Delta t} = kN$$

Michaelis–Menten equation

$$\frac{d[P]}{dt} = \frac{k_2[E]_0[S]}{K_M + [S]} \quad \text{where } K_M = \frac{k_2 + k_{-1}}{k_1}$$

molar mass

$$M = \frac{m}{n}$$

mole fraction

$$\text{mole fraction of } A = x_A = \frac{n_A}{n_{total}}$$

Nernst equation

$$E_{cell} = E^{\ominus}_{cell} - \frac{RT}{zF}\ln Q$$

NMR chemical shift

$$\delta = \frac{\text{shift in frequency of a signal from TMS (Hz)}}{\text{operating frequency of the spectometer (MHz)}}$$

partial pressure of a component of a gas mixture

$$p_A = x_A\, p_{total}$$

percentage by mass

$$\%\text{ by mass of element} = \frac{\text{mass of element present in the sample}}{\text{mass of whole sample}} \times 100\%$$

percentage yield

$$\text{percentage yield} = \frac{\text{actual yield}}{\text{theoretical yield}} \times 100\%$$

pH

$$pH = -\log[H_3O^+]$$
$$pOH = -\log[OH^-]$$
$$pH + pOH = pK_w = 14.00 \text{ (at } 25.0\,°C)$$

pK_a and pK_b, relationship

$$pK_a + pK_b = pK_w = 14.00 \text{ (at } 25.0\,°C)$$

radiocarbon dating

$$\ln\frac{r_0}{r_t} = (1.21 \times 10^{-4}\,\text{yr}^{-1})t$$

Raoult's law

$$p_{solution} = X_{solvent}\,p^*_{solvent}$$
$$p_{solution} = (1 - X_{solute})p^*_{solvent}$$
$$= p^*_{solvent} - X_{solute}\,p^*_{solvent}$$

Raoult's law for a two-component system:

$$p_{total} = X_A P^*_A + X_B p^*_B$$

rate of reaction for the general reaction:

$aA + bB \rightarrow cC + dD$

$$\text{rate of reaction} = -\frac{1}{a}\frac{d[A]}{dt} = -\frac{1}{b}\frac{d[B]}{dt} = \frac{1}{c}\frac{d[C]}{dt} = \frac{1}{d}\frac{d[D]}{dt}$$

reaction quotient for the solution-phase reaction:

$aA + bB \rightarrow cC + dD$

$$Q_c = \frac{[C]^c[D]^d}{[A]^a[B]^b}$$

Schrödinger equation

$$H\psi = E\psi$$

solubility product of a salt

$$M_aX_b(s) \rightleftharpoons aM^{b+}(aq) + bX^{a-}(aq)$$
$$K_{sp} = [M^{b+}]^a[X^{a-}]^b$$

specific heat

$$c = \frac{C}{m}$$

specific rotation

$$\text{specific rotation} = [\alpha]^T_\lambda = \frac{\text{observed rotation (degress)}}{\text{length (dm)} \times \text{concentration (g/mL)}}$$

speed of a light wave in a vacuum

$$c = \nu\lambda$$

van der Waals equation

$$\left(p + \frac{n^2a}{V^2}\right)(V - nb) = nRT$$

van't Hoff equation

$$\frac{d\ln K}{dT} = \frac{\Delta H^{\ominus}}{RT^2}$$

$$\ln K_{T_2} - \ln K_{T_1} = \frac{-\Delta_r H^{\ominus}}{R}\left(\frac{1}{T_2} - \frac{1}{T_1}\right)$$

van't Hoff equation for osmotic pressure

$$\Pi V = nRT$$

van't Hoff factor

$$i = \frac{(\Delta T_f)_{measured}}{(\Delta T_f)_{calculated\ as\ a\ nonelectrolyte}}$$

work done in the expansion/compression of a gas

$$w = -p\Delta V$$

REFERENCES

Section 1: General: Physical constants, units and symbols

1. a. Cohen, E. R., Cvitaš, T., Frey, J. G., Holmström B., Kuchitsu, K., Marquardt, R., Mills, I., Pavese, F., Quack, M., Stohner, J., Strauss, H L., Takami, M., Thor, A. J. *Quantities, Units and Symbols in Physical Chemistry*, 3rd Ed, RSC Publishing, 2007.
   b. *Definitions of the SI base units*. Website http://physics.nist.gov/cuu/Units/current.html.
2. a. Blackman, A., Bottle, S. E., Schmid, S., Mocerino, M., Wille, U. *Chemistry*, 2nd Ed, John Wiley & Sons Australia, 2012.
3. a. *CODATA Internationally Recommended Values of the Fundamental Physical Constants: 2010*. Website http://physics.nist.gov/cuu/Constants/index.html.
4. a. Cohen, E. R., Cvitaš, T., Frey, J. G., Holmström B., Kuchitsu, K., Marquardt, R., Mills, I., Pavese, F., Quack, M., Stohner, J., Strauss, H L., Takami, M., Thor, A. J. *Quantities, Units and Symbols in Physical Chemistry*, 3rd Ed, RSC Publishing, 2007.
   b. McGlashan, M. L., *Physico-Chemical Quantities and Units*, 2nd Ed, Royal Institute of Chemistry, London, 1971.
7. a. Cohen, E. R., Cvitaš, T., Frey, J. G., Holmström B., Kuchitsu, K., Marquardt, R., Mills, I., Pavese, F., Quack, M., Stohner, J., Strauss, H L., Takami, M., Thor, A. J. *Quantities, Units and Symbols in Physical Chemistry*, 3rd Ed, RSC Publishing, 2007.

Section 2: The elements

8. a. Emsley, J., *The Elements*, 3rd Ed, Oxford University Press, Oxford, 1998.
   b. Moore, C. E., *Ionization Potentials and Ionization Limits Derived from the Analyses of Optical Spectra*, National Standard Reference Data System NSRDS-NBS 34, Washington, 1971.
9. a. Ahrens, L. H., *Geochim. Cosmochim. Acta*, vol. 2, p. 155, 1952.
   b. Blackman, A., Bottle, S. E., Schmid, S., Mocerino, M., Wille, U. *Chemistry*, 2nd Ed, John Wiley & Sons Australia, 2012.
   c. Coplen, T. B. and Holden, N. E., *Chemistry International*, vol. 33, p. 10, 2011.
   d. Dean, J. A., *Lange's Handbook of Chemistry*, 13th and 14th Eds, McGraw-Hill, New York, 1985 and 1992.
   e. Emsley, J., *The Elements*, 3rd Ed, Oxford University Press, Oxford, 1998.
   f. Haynes, W. M., *CRC Handbook of Chemistry and Physics*, 93rd Ed, CRC Press, Boca Raton, 2012.
   g. Ho, C. Y., Powell, R. W., Liley, P. E., *Thermal Conductivity of the Elements: a Comprehensive Review, J. Phys. Chem. Ref. Data*, vol. 3, supplement 1, 1974.
   h. Kaye, G. W. C. and Laby, T. H., *Tables of Physical and Chemical Constants*, 15th Ed, Longmans, 1986.
   i. IUPAC, *Isotopic Composition of the Elements, 1997, Pure and Applied Chemistry*, vol. 70, p. 217, 1998.
   j. Shannon, R. D., *Acta Cryst.*, vol. A32, p. 751, 1976.
   k. Speight, J. G., *Lange's Handbook of Chemistry*, 16th Ed, McGraw-Hill, New York, 2005.
   l. Sutton, L. E., *Tables of Interatomic Distances and Configuration in Molecules and Ions*, Chemical Society Special Publications 11 and 18, London, 1958 and 1965.
   m. Wells, A. F., *Structural Inorganic Chemistry*, 5th Ed, Clarendon Press, Oxford, 1984.
   n. Wieser, M. E. and Coplen, T. B., *Atomic Weights of the Elements, 2009, Pure and Applied Chemistry*, vol. 83, p. 359. © IUPAC 2010.
10. a. *Summary of Radioactivity: Properties and Uses*. Website www.ausetute.com.au/nuclesum.html.
11. a. Allred, A. L., *J. Inorg. Nucl. Chem.*, vol. 17, p. 215, 1961.
    b. Haynes, W. M., *CRC Handbook of Chemistry and Physics*, 93rd Ed, CRC Press, Boca Raton, 2012.
    c. Pauling, L., *The Nature of the Chemical Bond*, 3rd Ed, Cornell University Press, Ithaca, 1960.
12. a. Chase, M. W., *NIST-JANAF Thermochemical Tables*, 4th Ed, *J. Phys. Chem. Ref. Data*, monograph 9, 1998.
    b. Cox, J. D., Wagman, D. D. and Medvedev, V. A., *CODATA Key Values for Thermodynamics*, Hemisphere, New York, 1989.
    c. Emsley, J., *The Elements*, 3rd Ed, Oxford University Press, Oxford, 1998.
    d. Haynes, W. M., *CRC Handbook of Chemistry and Physics*, 93rd Ed, CRC Press, Boca Raton, 2012.
    e. Kubaschewski, O. and Alcock, C. B., *Metallurgical Thermochemistry*, 5th Ed, Pergamon, Oxford, 1979.
    f. Nesmeyanov, A. N., *Vapour Pressure of the Elements*, Infosearch, London, 1963.

g. Wagman, D. D., Evans, W. H., Parker, V. B., Schumm, R. H., Halow, I., Bailey, S. M., Churney, K. L., Nuttall, R. L., *The NBS Tables of Chemical Thermodynamic Properties*, *J. Phys. Chem. Ref. Data*, vol. 11, suppl. 2, 1982.

13. a. Haynes, W. M., *CRC Handbook of Chemistry and Physics*, 93rd Ed, CRC Press, Boca Raton, 2012.

14. a. Lias, S. G., Bartmess, J. E., Liebman, J. F., *Gas Phase Ion and Neutral Thermochemistry*, *J. Phys. Chem. Ref. Data*, vol. 17, suppl. 1, 1988.

b. Martin, W. C., Hagar, L., Reader, J. and Sugar, S., *Ground Levels and Ionization Potentials for Lanthanide and Actinide Atoms and Ions, J. Phys. Chem. Ref. Data*, vol. 3, p. 771, 1974.

c. Martin, W. C., Zalubas, R. and Hagar, L., *Atomic Energy Levels: The Rare Earth Elements*, National Standard Reference Data System NSRDS-NBS 60, Washington, 1978.

d. Moore, C. E., *Ionization Potentials and Ionization Limits Derived from the Analyses of Optical Spectra*, National Standard Reference Data System NSRDS-NBS 34, Washington, 1971.

15. a. Haynes, W. M., *CRC Handbook of Chemistry and Physics*, 93rd Ed, CRC Press, Boca Raton, 2012.

Section 3: Inorganic compounds

16. a. Chase, M. W., *NIST-JANAF Thermochemical Tables*, 4th Ed, *J. Phys. Chem. Ref. Data*, monograph 9, 1998.

b. Connelly, N. G., Damhus, T., Hartshorn, R. M., Hutton, A. T., *Nomenclature of Inorganic Chemistry*, RSC Publishing, 2005.

c. Cox, J. D., Wagman, D. D. and Medvedev, V. A., *CODATA Key Values for Thermodynamics*, Hemisphere, New York, 1989.

d. Dean, J. A., *Lange's Handbook of Chemistry*, 13th and 14th Ed, McGraw-Hill, New York, 1985 and 1992.

e. Haynes, W. M., *CRC Handbook of Chemistry and Physics*, 93rd Ed, CRC Press, Boca Raton, 2012.

f. Kubaschewski, O. and Alcock, C. B., *Metallurgical Thermochemistry*, 5th Ed, Pergamon, Oxford, 1979.

g. Macintyre, J. E. (Ed), *Dictionary of Inorganic Compounds*, Chapman & Hall, London, 1992.

h. McClellan, A. L., *Tables of Experimental Dipole Moments*, Freeman, San Francisco, 1963, and Rahard, El Cerrito, 1974.

i. Nelson, R. D., Lide, D. R. and Maryott, A. A., *Selected Values of Electric Dipole Moments for Molecules in the Gas Phase*, National Standard Reference Data System NSRDS-NBS 10, Washington, 1967.

j. Pedley, J. B., *Computer Analysis of Thermochemical Data* (CATCH), University of Sussex, Brighton, 1972.

k. Seidell, A. and Linke, W. F., *Solubilities of Inorganic and Metal-Organic Compounds*, 4th Ed, Van Nostrand, Princeton, 1958 and 1965.

l. Speight, J. G., *Lange's Handbook of Chemistry*, 16th Ed, McGraw-Hill, New York, 2005.

m. Stephen, H. and Stephen, T., *Solubilities of Inorganic and Organic Compounds*, Pergamon, Oxford, 1963.

n. Wagman, D. D., Evans, W. H., Parker, V. B., Schumm, R. H., Halow, I., Bailey, S. M., Churney, K. L., Nuttall, R. L., *The NBS Tables of Chemical Thermodynamic Properties, J. Phys. Chem. Ref. Data*, vol. 11, suppl. 2, 1982.

o. Wells, A. F., *Structural Inorganic Chemistry*, 5th Ed, Clarendon Press, Oxford, 1984.

17. a. Wells, A. F., *Structural Inorganic Chemistry*, 5th Ed, Clarendon Press, Oxford, 1984.

18. a. Wells, A. F., *Structural Inorganic Chemistry*, 5th Ed, Clarendon Press, Oxford, 1984.

b. Sutton, L. E., *Tables of Interatomic Distances and Configuration in Molecules and Ions*, Chemical Society Special Publications 11 and 18, London, 1958 and 1965.

19. a. Sutton, L. E., *Tables of Interatomic Distances and Configuration in Molecules and Ions*, Chemical Society Special Publications 11 and 18, London, 1958 and 1965.

20. a. Cottrell, T. L., *The Strengths of Chemical Bonds*, 2nd edn, Butterworths, London, 1958.

b. Johnson, D. A., *Some Thermodynamic Aspects of Inorganic Chemistry*, 2nd edn, Cambridge University Press, Cambridge, 1982.

c. Ball, M. C. & Norbury, A. H., *Physical Data for Inorganic Chemists*, Longman, London, 1974.

21. a. Jenkins, H. D. B in Ellis, H., *Nuffield Advanced Science Book of Data*, Longman, London, 1984.

b. Jenkins, H. D. B in Lide, D. R., *CRC Handbook of Chemistry and Physics*, 73rd, 76th and 81st Eds, CRC Press, Boca Raton, 1992, 1995 and 2000.

c. Wagman, D. D., Evans, W. H., Parker, V. B., Schumm, R. H., Halow, I., Bailey, S. M., Churney, K. L., and Nuttall, R. L., *The NBS Tables of Chemical Thermodynamic Properties*, *J. Phys. Chem. Ref. Data*, vol. 11, suppl. 2, 1982.

22. a. Sillen, L. G. and Martell, A. E., *Stability Constants of Metal-Ion Complexes*, Chemical Society Special Publications 17 and 25, London, 1964 and 1971.
b. Martell, A. E. and Smith, R. M., *Critical Stability Constants, Volumes 1 to 6*, Plenum, New York, 1974–1989.
23. a. Hogfeldt, E., *Stability Constants of Metal-ion Complexes, Part A: Inorganic Ligands*, Pergamon, Oxford, 1982.
b. Sillen, L. G. and Martell, A. E., *Stability Constants of Metal-Ion Complexes*, Chemical Society Special Publications 17 and 25, London, 1964 and 1971.

Section 4: Organic compounds

24. a. Blackman, A., Bottle, S. E., Schmid, S., Mocerino, M., Wille, U. *Chemistry*, 2nd Ed, John Wiley & Sons Australia, 2012.
25. a. *A Guide to IUPAC Nomenclature of Organic Compounds: Recommendations 1993*, Blackwell, Oxford, 1993.
b. Cox, J. D. and Pilcher, G., *Thermochemistry of Organic and Organometallic Compounds*, Academic, London, 1970.
c. Daubert, T. E. and Danner, R. P., *Physical and Thermodynamic Properties of Pure Chemicals*: DIPPR Data Compilation, Hemisphere, 1989.
d. Haynes, W. M., *CRC Handbook of Chemistry and Physics*, 93rd Ed, CRC Press, Boca Raton, 2012.
e. Jones, J. C., *Uncertainties in the Flash Point of Dimethylether, Journal of Loss Prevention in the Process Industries*, vol. 14, p. 429, 2001.
f. Martell, A. E. and Smith, R. M., *Critical Stability Constants, Volumes 1 to 6*, Plenum, New York, 1974–1989.
g. McClellan, A. L., *Tables of Experimental Dipole Moments*, Freeman, San Francisco, 1963, and Rahard, El Cerrito, 1974.
h. Nelson, R. D., Lide, D. R. and Maryott, A. A., *Selected Values of Electric Dipole Moments for Molecules in the Gas Phase*, National Standard Reference Data System NSRDS-NBS 10, Washington, 1967.
i. Pedley, J. B., Naylor, R. D. and Kirby, S. P., *Thermochemical Data of Organic Compounds*, 2nd Ed, Chapman and Hall, London, 1986.
j. Sergeant, E. P. and Dempsey, B., *Ionisation Constants of Organic Acids in Aqueous Solution*, Pergamon, Oxford, 1979.
k. Sillen, L. G. and Martell, A. E., *Stability Constants of Metal-Ion Complexes*, Chemical Society Special Publications 17 and 25, London, 1964 and 1971.
l. Stull, D. R., Westrum, E. F. and Sinke, G. C., *The Chemical Thermodynamics of Organic Compounds*, Wiley, New York, 1969.
m. Wagman, D. D., Evans, W. H., Parker, V. B., Schumm, R. H., Halow, I., Bailey, S. M., Churney, K. L., and Nuttall, R. L., *The NBS Tables of Chemical Thermodynamic Properties, J. Phys. Chem. Ref. Data*, vol. 11, suppl. 2, 1982.
26. a. Domalski, E. S., *Heat Capacities and Entropies of Organic Compounds in the Condensed Phase, J. Phys. Chem. Ref. Data*, vol. 13, suppl. 1, 1984.
b. Fasman, G. D., *Practical Handbook of Biochemistry and Molecular Biology*, CRC Press, Boca Raton, 1989.
c. Miller, S. L. and Smith-Magowan, D., *The Thermodynamics of the Krebs Cycle and Related Compounds, J. Phys. Chem. Ref. Data*, vol. 19, p. 1049, 1990.
27. a. Vallat, A., *Good laboratory practice for HPLC*. Website http://www2.unine.ch/files/content/sites/saf/files/shared/documents/GoodLaboratoryPractice-HPLC.pdf.
29. a. Dean, J. A., *Lange's Handbook of Chemistry*, 13th and 14th Eds, McGraw-Hill, New York, 1985 and 1992.
b. Haynes, W. M., *CRC Handbook of Chemistry and Physics*, 93rd Ed, CRC Press, Boca Raton, 2012.
c. Kaye, G. W. C. and Laby, T. H., *Tables of Physical and Chemical Constants*, 15th Ed, Longmans, 1986.
30. a. Daubert, T. E. and Danner, R. P., *Physical and Thermodynamic Properties of Pure Chemicals*: DIPPR Data Compilation, Hemisphere, 1989.
b. Dean, J. A., *Lange's Handbook of Chemistry*, 13th and 14th Eds, McGraw-Hill, New York, 1985 and 1992.
c. Haynes, W. M., *CRC Handbook of Chemistry and Physics*, 93rd Ed, CRC Press, Boca Raton, 2012.
d. Kaye, G. W. C. and Laby, T. H., *Tables of Physical and Chemical Constants*, 15th Ed, Longmans, 1986.
e. Speight, J. G., *Lange's Handbook of Chemistry*, 16th Ed, McGraw-Hill, New York, 2005.

Section 5: Spectroscopic data

32. a. Williams, D. H. and Fleming, I., *Spectroscopic Methods in Organic Chemistry*, 4th Ed, McGraw-Hill, London, 1987.
33. a. Haynes, W. M., *CRC Handbook of Chemistry and Physics*, 93rd Ed, CRC Press, Boca Raton, 2012.

b. Williams, D. H. and Fleming, I., *Spectroscopic Methods in Organic Chemistry*, 4th Ed, McGraw-Hill, London, 1987.

34. a. Gottlieb, H. G., Kotlyar, V. C., Nudelman, A., *J. Org. Chem.*, vol. 62, p. 7512, 1997.

35. a. Gottlieb, H. G., Kotlyar, V. C., Nudelman, A., *J. Org. Chem.*, vol. 62, p. 7512, 1997.

36. a. Barron, A. R., *NMR properties of the elements*. Website http://cnx.org/content/m34632/latest/.

b. Drago, R. S., *Physical Methods for Chemists*, 2nd Ed, Saunders College Publishing, 1992.

37. a. Silverstein, R. M., Webster, F. X., Kiemle, D. J *Spectrometric Identification of Organic Compounds*, John Wiley & Sons, 7th Ed, p. 70, 2005.

Section 6: Properties of acids and bases

38. a. Martell, A. E. and Smith, R. M., *Critical Stability Constants, Volumes 1 to 6*, Plenum, New York, 1974–1989.

b. Perrin, D. D., *Ionisation Constants of Inorganic Acids and Bases in Aqueous Solution*, 2nd Ed, Pergamon, Oxford, 1982.

c. Sillen, L. G. and Martell, A. E., *Stability Constants of Metal-Ion Complexes*, Chemical Society Special Publications 17 and 25, London, 1964 and 1971.

39. a. Mendham, J. Denney, R. C. Barnes, J. D. Thomas, M. J. K., *Vogel's Textbook of Quantitative Chemical Analysis*, Prentice Hall, 6th Ed, 2000.

b. California State University, *Properties of commercial acids and bases*. Website http://www.csudh.edu/oliver/chemdata/acid-str.htm.

40. a. Haynes, W. M., *CRC Handbook of Chemistry and Physics*, 93rd Ed, CRC Press, Boca Raton, 2012.

41. a. Adapted from Dhanlal De Lloyd, with data from the Physical and Theoretical Laboratory, Oxford University.

b. Mendham, J. Denney, R. C. Barnes, J. D. Thomas, M. J. K., *Vogel's Textbook of Quantitative Chemical Analysis*, Prentice Hall, 6th Ed, 2000.

c. *Preparation of pH buffer solutions*. Website http://delloyd.50megs.com/moreinfo/buffers2.html.

42. a. *SigmaAldrich Buffer Reference Center*. Website http://www.sigmaaldrich.com/life-science/core-bioreagents/biological-buffers/learning-center/buffer-reference-center.html.

Section 7: Properties of aqueous solutions

43. a. Robinson, R. A. and Stokes, R. H., *Electrolyte Solutions*, 2nd Ed, Butterworths, London, 1959.

b. Washburn, E. W. and West, C. J., *International Critical Tables*, McGraw-Hill, New York, 1926.

44. a. Appel, R., Bartels, J., Eucken, A., Hellwege, K.-H., *Landolt-Börnstein: Elektrische Eigenschaften* 2, 2 Band, 7 Teil, 1960.

b. Hamer, W. J. and de Wane, H. J., *Electrolytic Conductance and the Conductances of the Halogen Acids in Water*, National Standard Reference Data System NSRDS-NBS 33, Washington, 1970.

c. Harned, H. S. and Owen, B. B., *The Physical Chemistry of Electrolyte Solutions*, 3rd Ed, Reinhold, New York, 1958.

45. a. Apple, R., Bartels, J., Eucken, A., Hellwege, K-H., *Landolt-Börnstein: Elektrische Eigenschaften* 2, 2 Band, 7 Teil, 1960.

b. Harned, H. S. and Owen, B. B., *The Physical Chemistry of Electrolyte Solutions*, 3rd Ed, Reinhold, New York, 1958.

c. Marsh, K. N. and Stokes, R. H., *Aust. J. Chem.*, vol. 17, p. 740, 1964.

d. Robinson, R. A. and Stokes, R. H., *Electrolyte Solutions*, 2nd Ed, Butterworths, London, 1959.

46. a. Dean, J. A., *Lange's Handbook of Chemistry*, 13th and 14th Eds, McGraw-Hill, New York, 1985 and 1992.

b. Seidell, A., *Solubilities of Organic Compounds*, 3rd Ed, Van Nostrand, New York, 1941.

c. Seidell, A. and Linke, W. F., *Solubilities of Inorganic and Metal-Organic Compounds*, 4th edn, Van Nostrand, Princeton, 1958 and 1965.

d. Stephen, H. and Stephen, T., *Solubilities of Inorganic and Organic Compounds*, Pergamon, Oxford, 1963.

47. a. Dean, J. A., *Lange's Handbook of Chemistry*, 13th and 14th Eds, McGraw-Hill, New York, 1985 and 1992.

b. Dorsey, N. E., *Properties of Ordinary Water Substance*, Reinhold, New York, 1940.

c. Speight, J. G., *Lange's Handbook of Chemistry*, 16th Ed, McGraw-Hill, New York, 2005.

d. Washburn, E. W. and West, C. J., *International Critical Tables*, McGraw-Hill, New York, 1926.

48. a. Timmermans, J., *The Physico-Chemical Constants of Binary Systems in Concentrated Solutions*, Interscience, New York, 1959 to 1960.

b. Washburn, E. W. and West, C. J., *International Critical Tables*, McGraw-Hill, New York, 1926.

Section 8: Electrochemistry

49. a. Bard, A. J., Parsons, R. and Jordan, J., *Standard Potentials in Aqueous Solution* (IUPAC), Marcel Dekker, New York, 1985.
    b. Bratsch, S. G., *Standard Electrode Potentials and Temperature Coefficients in Water at 298.15 K, J. Phys. Chem. Ref. Data*, vol. 18, p. 1, 1989.
    c. Charlot, G., Collumeau, A. and Marchon, M. J. C., *Selected Constants: Oxidation-Reduction Potentials of Inorganic Substances in Aqueous Solution*, Butterworths, London, 1971.
    d. Haynes, W. M., *CRC Handbook of Chemistry and Physics*, 93rd Ed, CRC Press, Boca Raton, 2012.
    e. Sillen, L. G. and Martell, A. E., *Stability Constants of Metal-Ion Complexes*, Chemical Society Special Publications 17 and 25, London, 1964 and 1971.
50. a. Mendham, J., Denney, R. C., Barnes, J. D., Thomas, M. J. K., *Vogel's Textbook of Quantitative Chemical Analysis*, Prentice Hall, 6th Ed, 2000.
    b. Research Solutions & Resources LLC, *Potentials of Common Reference Electrodes*. Website http://www.consultrsr.com/resources/ref/refpotls.htm#mse.
    c. Sensortechnik Meinsberg, *Meinsberg Electrodes*. Website http://www.meinsberger-electroden.de/en/ueberblick/pot.html.
51. a. Pavlishchuk, V. V., Addison, A. W., *Inorg. Chim. Acta* vol. 298, p. 97, 2000.

Section 9: Appendices

52. a. Safe Work Australia. Website http://www.safeworkaustralia.gov.au/sites/swa/whs-information/hazardous-chemicals/ghs/pages/ghs.
    b. Sigma Aldrich. Website http://www.sigmaaldrich.com/safety-center/globally-harmonized.html.

INDEX

References are to the table number and the page numbers.
Table numbers are in **bold** type and the page numbers in *italic* type.